CLIMATE CHANGE AND ITS CAUSES, EFFECTS AND PREDICTION SERIES

THE ROLE OF AGRICULTURE IN CARBON CAPTURE AND CLIMATE CHANGE

Climate Change and its Causes, Effects and Prediction Series

Global Climate Change
Horace M. Karling (Editor)
2001. ISBN: 1-56072-999-6

Global Climate Change Revisited
Harace B. Karling (Editor)
2007. ISBN: 1-59454-039-X

Climate Change Research Progress
Lawrence N. Peretz (Editor)
2008. ISBN: 1-60021-998-5

Climate Change: Financial Risks
United States Government Accountability Office
2008. ISBN: 978-1-60456-488-4

Post-Kyoto: Designing the Next International Climate Change Protocol
Matthew Clarke
2008. ISBN: 978-1-60456-840-0

Post-Kyoto: Designing the Next International Climate Change Protocol
Matthew Clarke
2008. ISBN: 978-1-61668-101-2 (Online Book)

The Effects of Climate Change on Agriculture, Land Resources, Biodiversity in the United States
Peter Backlund, Anthony Janeto and David Schimel
2009. ISBN: 978-1-60456-989-6

Economics of Policy Options to Address Climate Change
Gregory N. Bartos
2009. ISBN: 978-1-60692-116-6

Climate Variability, Modeling Tools and Agricultural Decision-Making
Angel Utset (Editor)
2009. ISBN: 978-1-60692-703-8

Climate Variability, Modeling Tools and Agricultural Decision-Making
Angel Utset (Editor)
2009. ISBN: 978-1-60876-791-5 (Online Book)

Emissions Trading: Lessons Learned from the European Union and Kyoto Protocol Climate Change Programs
Ervin Nagy and Gisella Varga (Editors)
2009. ISBN: 978-1-60741-194-9

Constructing Climate Change Legislation: Background and Issues
Gerald P. Overhauser (Editor)
2009. ISBN: 978-1-60692-986-5

Designing Greenhouse Gas Reduction and Regulatory Systems
Sonja Enden (Editor)
2009. ISBM: 978-1-60741-195-6

Global Climate Change: International Perspectives and Responses
Elias D'Angelo (Editor)
2009. ISBN: 978-1-60741-233-5

Disputing Global Warming
Anton Horvath and Boris Molnar (Editors)
2009. ISBN: 978-1-60741-235-9

Disputing Global Warming
Anton Horvath and Boris Molnar (Editors)
2009. ISBN: 978-1-60876-503-4 (Online Book)

Focus on Climate Change and Health
Viroj Wiwanitkit
2009. ISBN: 978-1-60741-247-2

Land Use and Climate Change
Suresh C. Rai
2009. ISBN: 978-1-60741-362-2

Lightning in the Tropics: From a Source of Fire to a Monitoring System of Climatic Changes
Osmar Pinto, Jr.
2009. ISBN: 978-1-60741-764-4

Global Change and Forestry: Economic and Policy Impacts and Responses
Jianbang Gan, Stephen Grado and Ian A. Munn (Editors)
2010. ISBN: 978-1-60876-262-0

Carbon Tax and Cap-and-Trade Tools: Market-Based Approaches for Controlling Greenhouse Gases
Nelson E. Burney (Editor)
2010. ISBN: 978-1-60876-137-1

Carbon Offsets: Examining their Role in Greenhouse Gas Reduction
Karen T. Morningstar (Editor)
2010. ISBN: 978-1-60741-444-5

The Science of Climate Change and Policy Implications
Kyle S. Hartzell (Editor)
2010. ISBN: 978-1-60741-448-3

Carbon Sequestration: Methods, Modeling and Impacts
Elke Hoch and Siegbert Grunwald (Editors)
2010. ISBN: 978-1-60741-498-8

The Role of Agriculture in Carbon Capture and Climate Change
Arvid Bjurstrom (Editor)
2010. ISBN: 978-1-60741-445-2

CLIMATE CHANGE AND ITS CAUSES, EFFECTS AND PREDICTION SERIES

THE ROLE OF AGRICULTURE IN CARBON CAPTURE AND CLIMATE CHANGE

ARVID BJURSTROM
EDITOR

Nova Science Publishers, Inc.
New York

For permission to use material from this book please contact us:
Telephone 631-231-7269; Fax 631-231-8175
Web Site: http://www.novapublishers.com

LIBRARY OF CONGRESS CATALOGING-IN-PUBLICATION DATA

The role of agriculture in carbon capture and climate change / editor: Arvid Bjurstrom.
p. cm.
Includes index.
ISBN 978-1-60741-445-2 (softcover)
1. Climatic changes--United States. 2. Carbon sequestration--United States. I. Bjurstrom, Arvid.
QC903.2.U6R65 2010
363.738'74--dc22
2009038420

Published by Nova Science Publishers, Inc. ✝ New York

CONTENTS

PREFACE

The agriculture sector is a source of greenhouse gas (GHG) emissions, which many scientists agree are contributing to observed climate change. Agriculture is also a "sink" for sequestering carbon, which might offset GHG emissions by capturing and storing carbon in agricultural soils. Emissions from agricultural activities account for 6%-8% of all GHG emissions in the United States. Carbon captured and stored in U.S. agricultural soils partially offsets these emissions, sequestering about one-tenth of the emissions generated by the agriculture sector, but less than 1% of all U.S. emissions annually. Congress is considering a range of climate change policy options, including GHG emission reduction programs that would either mandate or authorize a cap-and-trade program to reduce GHG emissions. In general, the current legislative proposals would not require emissions in the agriculture and forestry sectors. Many GHG proposals, however, would allow farmers and landowners to receive emissions allowances (or credits) and/or generate carbon offsets, which could be sold to facilities covered by a cap-and-trade program. This book highlights the effects that agriculture has on carbon capture and climate change.This book consists of public documents which have been located, gathered, combined, reformatted, and enhanced with a subject index, selectively edited and bound to provide easy access.

Chapter 1 - This report summarizes some of the domestic agriculture and forestry provisions in the Lieberman-Warner Climate Security Act of 2008 (S. 3036, formerly S. 2191), as ordered reported out of the Senate Committee on Environment and Public Works in December 2007. The bill directs the Administrator of the U.S. Environmental Protection Agency to establish a program to decrease greenhouse gas (GHG) emissions. The bill's cap-and-trade framework establishes a tradeable allowance system that includes a combination of auctions and free allocation of tradeable allowances. As part of this overall framework, S. 3036 includes three design mechanisms that may provide financial incentives to encourage land-based agricultural and forestry activities. These include provisions on carbon offsets, set-aside allowances, and auction proceeds.

Chapter 2 - The agriculture sector is a source of greenhouse gas (GHG) emissions, which many scientists agree are contributing to observed climate change. Agriculture is also a "sink" for sequestering carbon, which might offset GHG emissions by capturing and storing carbon in agricultural soils. The two key types of GHG emissions associated with agricultural activities are methane (CH_4) and nitrous oxide (N_2O). Agricultural sources of CH_4 emissions mostly occur as part of the natural digestive process of animals and manure management at livestock operations; sources of N_2O emissions are associated with soil management and fertilizer use on croplands. This report describes these emissions on a carbon-equivalent basis to illustrate agriculture's contribution to total national GHG emissions and to contrast emissions against estimates of sequestered carbon.

Chapter 3 - Proposals to reduce emissions of carbon dioxide and other greenhouse gases often include the use of forestry and agricultural practices and lands for carbon sequestration. However, uncertainty about the accuracy of measuring carbon from these activities has led some to question this potential. Basic approaches for measuring forest and agricultural carbon include on-site measurement; indirect measurement from off-site tools; and estimation using models or inferences. Because of challenges associated with balancing the cost and accuracy of these measurement tools, any practicable system for measuring forest and agricultural carbon might require a mix of these approaches.

In: The Role of Agriculture in Carbon Capture... ISBN: 978-1-60741-445-2
Editor: Arvid Bjurstrom

Chapter 1

AGRICULTURE AND FORESTRY PROVISIONS IN CLIMATE CHANGE LEGISLATION*

Renée Johnson
Agricultural Policy Resources, Science and Industry Division

SUMMARY

This report summarizes some of the domestic agriculture and forestry provisions in the Lieberman-Warner Climate Security Act of 2008 (S. 3036, formerly S. 2191), as ordered reported out of the Senate Committee on Environment and Public Works in December 2007. The bill directs the Administrator of the U.S. Environmental Protection Agency to establish a program to decrease greenhouse gas (GHG) emissions. The bill's cap-and-trade framework establishes a tradeable allowance system that includes a combination of auctions and free allocation of tradeable allowances. As part of this overall framework, S. 3036 includes three design mechanisms that may provide financial incentives to encourage land-based agricultural and forestry activities. These include provisions on carbon offsets, set-aside allowances, and auction proceeds.

In the 110th Congress, several proposals have been introduced that would either mandate or authorize a cap-and-trade program to reduce greenhouse gas (GHG) emissions. A cap-and-trade program provides a market-based policy tool

---

* This is an edited, reformatted and augmented version of a CRS Report for Congress publication dated June 2008.

for reducing emissions by setting a cap, or maximum emissions limit, for certain industries. Sources covered by the cap can choose to reduce their own emissions, or can choose to buy emission credits that are generated from reductions made by other sources. This type of market-based approach to GHG reductions and trading would be similar to the acid rain reduction program established by the 1990 Clean Air Act Amendments.[1]

Among the cap-and-trade proposals introduced to date, none includes the agriculture sector as a covered industry subject to emission reductions under the cap.[2] In part, this may reflect the general consensus, as stated by the House Energy and Commerce Committee, that GHG "emissions from the agriculture sector generally do not lend themselves to regulation under a cap-and-trade program," given the "large number of sources with small individual emissions that would be impractical to measure."[3]

However, several of the cap-and-trade proposals do incorporate the agriculture and forestry sectors either as a source of carbon offsets[4] or as a recipient of set-aside allowances.[5] Some bills also specify that the proceeds from auctioned allowances be used to promote certain objectives, which could further encourage farmland conservation and bio-energy technologies and practices, among other activities.

Inclusion of such provisions in the broader cap-and-trade proposals could benefit the U.S. agriculture and forestry sectors. For example, the offset and allowance provisions would allow farmers and landowners to participate in the emerging market by granting them use of allowancesand credits for sequestration and/or emission reduction activities. These allowances and credits could be sold to regulated facilities (e.g., power plants) covered by a cap-and-trade program to meet their emission reduction obligations. The proceeds from the sale of these allowances and credits, as well as proceeds from auctions that fund technology deployment, are intended to further promote and support activities in the agriculture or forestry sectors that aim to reduce, avoid, or sequester emissions.

In the Senate, for example, a bill ordered reported by the Senate Committee on Environment and Public Works (EPW) in December 2007, the Lieberman-Warner Climate Security Act of 2008 (S. 3036, formerly S. 2191),[6] contains several agriculture-based provisions. A summary of these provisions is provided below. Overall, S. 3036 directs the Administrator of the U.S. Environmental Protection Agency to establish a program to decrease GHG emissions under a cap-and-trade framework.[7]

AGRICULTURE AND FORESTRY PROVISIONS

The cap-and-trade framework outlined in S. 3036 establishes a tradeable allowance system that includes a combination of auctions and free allocation of tradeable allowances. As part of this overall framework, S. 3036 includes three design mechanisms that may provide financial incentives to encourage land-based agricultural and forestry activities: carbon offsets, set-aside allowances, and auction proceeds. In this context, a carbon offset refers to a measurable avoidance, reduction, or sequestration of CO_2 or other GHG emissions, expressed in carbon-equivalent terms. A set-aside allowance refers to a set percentage of available allowances under the overall emissions cap that is allocated to non-regulated entities, in this case domestic agriculture and forestry entities. For auction proceeds, this refers to the set percentage that is allocated for use to carry out the cellulosic biomass ethanol technology deployment program.[8]

Offsets

Title II, Subtitle D ("Offsets"), of S. 3036 provides for agriculture and forestry offset projects. The agriculture and forestry provisions in this subtitle cover farmer outreach (Sec. 2401), establishment of a domestic offset program (Sec. 2402), eligible offset project types (Sec. 2403), project initiation and approval (Sec. 2404), offset verification and issuance of allowances (Sec. 2405), tracking of reversals for sequestration projects (Sec. 2406), examination and auditing of offset allowances (Sec. 2407), timing and the provision of offset allowances (Sec. 2408), offset registry (Sec. 2409), certain environmental considerations (Sec. 2410), program review (Sec. 2411), and retail carbon offset requirements (Sec. 2412). The text box below shows the types of eligible agriculture and forestryoffset projects listed in Section 2403 of S. 3036, which includes these listed practices or combinations of agricultural conservation practices.

In general, these types of conservation and farmland management practices are among existing agricultural and forestry programs that are administered at both the federal and state levels. Many of these practices are provided for as part of existing conservation, forestry, energy, and rural development programs under the 2008 farm bill (P.L. 110-234). These include conservation programs provided for in Title II of the farm bill, such as the Conservation Reserve Program, the Grasslands Reserve Program, the Environmental Quality Incentives Program, and theConservation Stewardship Program, among others. These programs provide technical assistance and either cost-sharing or easement payments that, in addition to accomplishing other environmental objectives, generally encourage land retirement or the types of agricultural practices that can reduce GHG emissions and/or sequester carbon (**Table 1**). Other farm bill programs in the Energy (Title IX) and Rural Development (Title VI) titles authorize loans, loan guarantees, and grants for energy efficiency and renewable energy systems, including anaerobic digesters. For more information, see CRS Report RL33898, *Climate Change: The Role of the U.S. Agriculture Sector*, by Renée Johnson.

Table 1. Current Conservation and Land Management Practices

USDA Program	Conservation Practice and Land Management	General Objectives	Objectives for Climate Change
EQIP, CSP, AMA	Conservation tillage and reduced field pass intensity	Improve soil/water/air quality. Reduces soil erosion/fuel use.	Sequestration, emission reduction
	Crop diversity through crop rotations and cover cropping	Reduce erosion/water needs. Improves soil/water quality.	Sequestration
	Efficient nutrient (nitrogen) management, fertilizer application	Improves water quality. Saves expenses, time, and labor.	Sequestration, emission reduction
	Improved soil management and soil erosion controls	Improve soil/water/air quality.	Sequestration, emission reduction
EQIP CSP AMA Othera	Manure management (e.g., storage/containment, anaerobic digestion and methane recovery)	Improve soil/water/air quality. On-farm fuel cost-savings. Alternative income source. Nutrients for crops.	Emission reduction
EQIP CSP AMA	Feed management (e.g., raise feed efficiency, dietary supplements) Rangeland management (e.g., rotational grazing, improved forage)	Improve water/air quality. More efficient use of feed. Reduce water require-ments. Help withstand drought. Raise grassland productivity.	Emission reduction Sequestration, emission reduction
EQIP CSP AMA WHIP	Windbreaks for crops and livestock, egetative/riparian buffers, grassed waterways, setbacks, etc.	Improve crop/livestock protection and wildlife habitat. Alternative income source (e.g., hunting fees).	Sequestration, emission reduction

Table 1. (Continued)

USDA Program	Conservation Practice and Land Management	General Objectives	Objectives for Climate Change
FLEP EQIP CSP AMA	Agroforestry / silvopasture with rotational grazing and improved forage	Provide income from grazing and wood products.	Sequestration, emission reduction
CRP WRP GRP FPP	Land management, including retirement, conversion, restoration (cropland, grasslands, wetlands, open space)	Improve soil/water/air quality.	Sequestration
EQIP CSP AMA Othera	Energy efficiency/conservation	Improve soil/water/air quality. Cost-savings.	Emission reduction
	Biofuel substitution and renewable energy use	Improve soil/water/air quality. On-farm fuel cost-savings. Alternative income source.	Emission reduction

Source: Compiled by CRS staff from USDA and EPA information. Listed programs: Conservation Reserve Program (CRP), Wetlands Reserve Program (WRP), Grasslands Reserve Program (GRP), Farmland Protection Program (FPP), Environmental Quality Incentives Program (EQIP), Conservation Stewardship Program (CSP), Agricultural Management Assistance (AMA), Wildlife Habitat Incentives Program (WHIP), and Forest Land Enhancement Program (FLEP).

a. Renewable energy projects receive additional program funding in the 2002 farm bill under Title IX (Energy) and Title VI (Rural Development), as well as other federal and state programs.

ELIGIBLE AGRICULTURAL AND FORESTRY OFFSET PROJECTS (S. 3036, SEC. 2403)

Agricultural/Rangeland Sequestration and Management Practices

- altered tillage practices
- winter cover cropping, continuous cropping, and other ways to increase biomass returned (other than planting followed by fallowing)
- conversion of cropland, rangeland, or grassland (with conditions)
- reduction of nitrogen fertilizer use or increase in nitrogen efficiency
- reduction in the frequency and duration of flooding of rice paddies
- reduction in carbon emissions from organic soils

Land Use Change and Forestry Activities (changes in carbon stocks)

- limited to afforestation or reforestation of acreage (not currently forested)
- forest management resulting in an increase in forest stand volume

Manure Management and Disposal

- waste aeration
- methane capture and combustion

Other Terrestrial Offset Practices Identified by USDA

- capture or reduction of non-covered fugitive emissions
- methane capture and combustion at nonagricultural facilities
- other actions that result in GHG emissions avoidance or reduction

Set-Aside Allowances

Title III, Subtitle G ("Domestic Agriculture and Forestry"), of S. 3036 directly allocates 5% of the overall emissions allowances to domestic agriculture and forestry entities (Sec. 3701). This could provide a sizeable benefit to U.S. producers. Overall, the proposal starts off with 5.8 billion emissions allowances for CY2012, which phases down to 1.7 billion emissions allowances for CY2050 (Sec. 1201). A 5% set-aside for domestic agriculture and forestry entities could give these sectors a significant part of this emerging market — between 290 million and 90 million emissions allowances for qualifying entities, depending on the year.

The agriculture and forestry provisions in this subtitle cover allocation (Sec. 3701), research (Sec. 3702), and distribution (Sec. 3703). The subtitle does not specify the types practices that would be applicable. However, it does state that emissions reduction and increases in carbon sequestration in the agriculture and forestry sectors should be "real, verifiable, additional, permanent, and enforceable" (Sec. 3701); it also specifies the need for reductions of both nitrous oxide emissions through soil management, and methane emissions through feed and manure management (Sec. 3702(a)).

This provision indirectly relates to a new USDA conservation provision that was included in the 2008 farm bill. This provision would facilitate the market development of environmental services from the agriculture and forestry sectors, including carbon storage and tradeable credits, by addressing measurement, quantification, verification, and enforcement issues, among other related issues. For information, see CRS Report RL34042, *Environmental Services Markets: Farm Bill Proposal*, by Renée Johnson.

Auction Proceeds

Title IV, Subtitle D ("Energy Technology Deployment"), of S. 3036 specifies that 6% of auction proceeds be used to carry out a variety of projects to promote cellulosic biomass ethanol technology deployment (Sec. 4401, Sec. 4404). This provision calls for the use of producer incentives, such as loan guarantees and production payments, to promote the construction of production facilities and supporting infrastructure for cellulosic biomass. This could benefit U.S. agriculture and forestry producers that produce transportation fuels from cellulosic biomass using different feedstocks. This subtitle does not specify the types of practices that would be applicable.

Considerations for Congress

Many see the involvement of the agriculture and forestry sectors in a climate change mitigation strategy as an opportunity to further encourage farmers and landowners to make environmental improvements on their land and to transition to more sustainable production practices. Nevertheless, inclusion of the agriculture and forestry sectors in a cap-and-trade program has remained controversial since the Kyoto Protocol negotiations.[9] During those negotiations, there was marked disagreement among countries and interest groups, arguing either for or against the inclusion of offsets from the agriculture and forestry sectors.[10] The text box below lists some of the primary areas of concern regarding agriculture and forestry offsets and allowances. The EU's GHG emission program, the Emission Trading System (ETS), which was established in 2005, does not provide for agricultural or forestry projects and activities. Among the reasons are (1) pragmatic concerns regarding measurement and verification, given the sheer number of farmers and landowners, and (2) ideological concerns about granting too much flexibility in how emission reductions are met, which could undermine overall program goals.[11] For a more detailed discussion of these issues, see CRS Report RL34241, *Voluntary Carbon Offsets: Overview and Assessment*, by Jonathan L. Ramseur, and CRS Report RL33898, *Climate Change: The Role of the U.S. Agriculture Sector*, by Renée Johnson.

AGRICULTURAL/FORESTRY OFFSETS AND ALLOWANCES: AREAS OF CONCERN

- ***Permanence/Duration*** — land uses can change over time (e.g., forest lands to urban development, natural events such as fires or pests);
- ***Measurement/Accounting*** — measuring biological sequestration is difficult and estimates can vary, and actual emission reduction/sequestration depends on site-specific factors (e.g., location, climate, soil type, crop/vegetation, tillage practices, management);
- ***Additionality*** — some activities generating offsets would have occurred anyway under a pre-existing program or practice, and may not go beyond business as usual (BAU); reductions may be double-counted or attributable to other environmental goal/ programs;
- ***Effectiveness*** — the success of the mitigation practice depends on the type of practice, how well it is implemented and managed by the farmer or landowner, and the length of time the practice is undertaken; and
- ***Leakage*** — reductions in one place could result in additional emissions elsewhere.

End Notes

[1] For more information about the GHG legislative proposals and the carbon offset provisions in these bills, see CRS Report RL33846, *Greenhouse Gas Reduction: Cap-and-Trade Bills in the 110th Congress*, by Larry Parker and Brent D. Yacobucci; and CRS Report RL34067, *Climate Change Legislation in the 110th Congress*, by Jonathan L. Ramseur and Brent D. Yacobucci.

[2] Some GHG bills give authority to the U.S. Environmental Protection Agency to determine covered entities, which could potentially expand the types and number of entities covered.

[3] Committee on Energyand Commerce, "Climate Change Legislation Design White Paper: Scope of a Cap-and-Trade Program," prepared by committee staff, October 2007, available at [http://energycommerce.house.gov/Climate_Change/White_Paper.100307.pdf].

[4] Among the GHG bills that provide for agriculture and/or forestry offsets are S. 3036 (Lieberman/Warner), S. 280 (McCain/Lieberman), S. 317 (Feinstein), S. 1168 (Alexander/ Lieberman), S. 1177 (Carper), S. 1766 (Bingaman/Specter), and H.R. 620 (Olver).

[5] Primarily S. 3036 and also S. 1766 (Bingaman/Specter).

[6] On June 2, 2008, the Senate invoked cloture on S. 3036 and will proceed to debate the bill under unanimous-consent agreement.

[7] This analysis is based on legislative text in S. 3036 as of May 20, 2008, and does not include changes from possible amendments that may be considered by the Senate.

[8] In carbon market trading, an offset is a certificate representing the reduction of the equivalence of one metric ton of carbon dioxide emissions, the principal greenhouse gas. Offsets generally fall within the categories of biological sequestration, renewable energy, energy efficiency, and non-CO_2 greenhouse gas emissions reductions. For more information on allowances and auction proceeds in current GHG bills, see *Allocations for Carbon Allowances and Auctions under S. 2191*, by Brent D. Yacobucci (CRS general distribution memorandum).

[9] See, for example, E. Boyd, E. Corbera, B. Kjellén, M. Guitiérrez, and M. Estrada, "The Politics of 'Sinks' and the CDM: A Process Tracing of the UNFCCC Negotiations (pre-Kyoto to COP9)," Feb. 2007, draft submitted for *International Environmental Agreements;* also see two articles in *Nature*, no. 6812, Nov. 2000, "Deadlock in the Hague, but Hope Remains for Spring Climate Deal," and "Critical Politics of Carbon Sinks."

[10] Commonly referred to as "land use, land use change, forestry," or abbreviated as LULUCF.

[11] Comments and presentation by Michael Grubb, Chief Economist of the Carbon Trust, during a Congressional staff briefing, February 29, 2008. Although private parties subject to the ETS cap cannot purchase LULUCF offsets, EU governments can purchase eligible LULUCF offsets — i.e., from afforestation or reforestation projects — up to 1% of their state's base year (1990) emissions each year (See European Union Directive 2004/101/EC, October 27, 2004; Kyoto Protocol, Decision 17/CP.7, November 2001). The World Bank reported that global transactions of LULUCF offsets have only accounted for 6% of this allowable limit.

In: The Role of Agriculture in Carbon Capture... ISBN: 978-1-60741-445-2
Editor: Arvid Bjurstrom

Chapter 2

CLIMATE CHANGE: THE ROLE OF THE U.S. AGRICULTURE SECTOR*

Renée Johnson
Agricultural Policy Resources, Science and Industry Division

SUMMARY

The agriculture sector is a source of greenhouse gas (GHG) emissions, which many scientists agree are contributing to observed climate change. Agriculture is also a "sink" for sequestering carbon, which might offset GHG emissions by capturing and storing carbon in agricultural soils. The two key types of GHG emissions associated with agricultural activities are methane (CH_4) and nitrous oxide (N_2O). Agricultural sources of CH_4 emissions mostly occur as part of the natural digestive process of animals and manure management at livestock operations; sources of N_2O emissions are associated with soil management and fertilizer use on croplands. This report describes these emissions on a carbon-equivalent basis to illustrate agriculture's contribution to total national GHG emissions and to contrast emissions against estimates of sequestered carbon.

Emissions from agricultural activities account for 6%-8% of all GHG emissions in the United States. Carbon captured and stored in U.S. agricultural soils partially offsets these emissions, sequestering about one-tenth of the

---

* This is an edited, reformatted and augmented version of a CRS Report for Congress publication dated June 2008.

emissions generated by the agriculture sector, but less than 1% of all U.S. emissions annually. Emissions and sinks discussed in this report are those associated with agricultural production only. Emissions associated with on-farm energy use or with food processing or distribution, and carbon uptake on forested lands or open areas that might be affiliated with the farming sector, are outside the scope of this report.

Most land management and farm conservation practices can help reduce GHG emissions and/or sequester carbon, including land retirement, conservation tillage, soil management, and manure and animal feed management, among other practices. Many of these practices are encouraged under most existing voluntary federal and state agricultural programs that provide cost-sharing and technical assistance to farmers, predominantly for other production or environmental purposes. However, uncertainties are associated with implementingthese types of practices depending on site-specific conditions, the type of practice, how well it is implemented, the length of time a practice is undertaken, and available funding, among other factors. Despite these considerations, the potential to reduce emissions and sequester carbon on agricultural lands is reportedly much greater than current rates.

Congress is considering a range of climate change policy options, including GHG emission reduction programs that would either mandate or authorize a cap-and-trade program to reduce GHG emissions. In general, the current legislative proposals would not require emission reductions in the agriculture and forestry sectors. Many GHG proposals, however, would allow farmers and landowners to receive emissions allowances (or credits) and/or generate carbon offsets, which could be sold to facilities covered by a cap-and-trade program. In addition, the enacted 2008 farm bill includes provisions that could expand the scope of existing land-based conservation and other farm bill programs by providing incentives to encourage farmers and landowners to sequester carbon and reduce emissions associated with climate change, adopt energy efficiency measures, produce renewable energy feedstocks, and participate in markets for carbon storage.

The debate in Congress over whether and how to address possible future climate change is intensifying. Often, the role of the U.S. agriculture sector is invoked in this debate. Agriculture is a source of greenhouse gas (GHG) emissions, which many scientists agree are contributing to observed climate change. Agriculture is also a “sink” for sequestering carbon, which partly offsets these emissions. Carbon sequestration (the capture and storage of carbon) in agricultural soils can be an important component of a climate change mitigation strategy, limiting the release of carbon from the soil to the atmosphere.

Congress is considering a range of climate change policy options, including GHG emission reduction programs that would either mandate or authorize a cap-and-trade program to reduce GHG emissions. In general, the current legislative proposals would not require emission reductions in the agriculture and forestry sectors. However, some of these proposals would allow farmers and landowners to generate offsets in support of a cap-and-trade program. Other proposals would give farmers and landowners a share of available allowances (or credits) for sequestration and/or emission reduction activities. These offsets and allowances could be sold to facilities (e.g., power plants) covered by a cap-and-trade program. Some bills also specify that the proceeds from auctioned allowances be used to promote certain activities, including farmland conservation and developing bio-energy technologies.

In addition, the omnibus 2008 farm bill (Food, Conservation, and Energy Act of 2008, P.L. 110-246) could expand the scope of existing farm and forestry conservation programs in ways that could more broadly encompass certain aspects of these climate change initiatives. The bill provides incentives to encourage farmers and landowners to sequester carbon and reduce emissions associated with climate change, as well as produce renewable energy feedstocks. The bill also contains a new provision that will facilitate the participation of the agriculture and forestry sectors in emerging environmental services markets, focusing first on carbon storage.

This report is organized in three parts. First, it discusses the extent of GHG emissions associated with the U.S. agriculture sector, and cites current and potential estimates for U.S. agricultural soils to sequester carbon and partly offset national GHG emissions. Second, the report describes the types of land management and farm conservation practices that can reduce GHG emissions and/or sequester carbon in agricultural soils, highlighting those practices that are currently promoted under existing voluntary federal agricultural programs. The Appendix provides a summary primer of the key background information presented in these first two sections. Finally, the report describes ongoing legislative action within both the climate change and farm bill debates, and discusses the types of questions that may be raised regarding the roleof theU.S. agriculturesector in the broader climate change debate.

This report does not address the potential effects of global climate change on U.S. agricultural production. Such effects may arise because of increased climate variability and incidence of global environmental hazards, such as drought and/or flooding, pests, weeds, and diseases, or temperature and precipitation changes that might cause locational shifts in where and how agricultural crops are produced.[1]

This report also does not address how ongoing or anticipated initiatives to promote U.S. bioenergy production may effect efforts to reduce GHG emissions and/or sequester carbon, such as by promoting more intensive feedstock production and by encouraging fewer crop rotations and planting area setbacks, which could both raise emissions and reduce carbon uptake.[2]

AGRICULTURAL EMISSIONS AND SINKS

Agriculture is a both a source and a sink of greenhouse gases, generating emissions that enter the atmosphere and removing carbon dioxide (CO_2) from the atmosphere through photosynthesis and storing it in vegetation and soils (a process known as sequestration). Sequestration in farmland soils partially offsets agricultural emissions. Despite this offset, however, the U.S. agriculture sector remains a net source of GHG emissions.

Source of National Estimates

Estimates of GHG emissions and sinks for the U.S. agriculture sector presented in this report are the official U.S. estimates of national GHG emissions and carbon uptake, as published annually by the U.S. Environmental Protection Agency (EPA) in its *Inventory of U.S. Greenhouse Gas Emissions and Sinks*.[3] EPA's *Inventory* data reflect annual national emissions by sector and fuel, including estimates for the agriculture and forestry sectors. EPA's estimates rely on data and information from the U.S. Department of Agriculture (USDA), the Department of Energy, the Department of Transportation, the Department of Defense, and other federal departments. The EPA-published data are rigorously and openly peer reviewed through formal interagency and public reviews involving federal, state, and local government agencies, as well as private and international organizations. For the agriculture and forestry sectors, USDA publishes a supplement to EPA's *Inventory*, which builds on much of the same data and information, but in some cases provides a more detailed breakout by individual states and sources.[4]

In this CRS report, emissions from agricultural activities are aggregated in terms of carbon dioxide or CO_2-equivalents, and expressed as million metric tons ($MMTCO_2$-Eq.).[5] This aggregation is intended to illustrate agriculture's contribution to national GHG emissions and to contrast emissions against estimates of sequestered carbon.

Agricultural Emissions

Direct GHG Emissions

The types of GHG emissions associated with agricultural activities are methane (CH_4) and nitrous oxide (N_2O), which are two of the key gases that contribute to GHG emissions.[6] These gases are significant contributors to atmospheric warming and have a greater effect warming than the same mass of CO_2.[7]

Agricultural sources of CH_4 emissions mostly occur as part of the natural digestive process of animals and manure management in U.S. livestock operations. Sources of N_2O emissions are mostly associated with soil management and commercial fertilizer and manure use on U.S. croplands, as well as production of nitrogen-fixing crops.[8] Emissions of N_2O from agricultural sources account for about two-thirds of all reported agricultural emissions; emissions of CH_4account for about one-third of all reported emissions. Across all economic sectors, the U.S. agriculture sector was the leading source of N_2O emissions (80%) and a major source of CH_4 emissions (30%) in 2005.[9]

Other Types of Emissions

Agricultural activities may also emit other indirect greenhouse gases, such as carbon monoxide, nitrogen oxides, and volatile organic compounds from field burning of agricultural residues.[10] These emissions are not included in EPA's annual *Inventory* estimates because they contribute only indirectly to climate change by influencing tropospheric ozone, which is a greenhouse gas. Agricultural activities may also release other types of air emissions, some of which are regulated under the federal Clean Air Act, including ammonia, volatile organic compounds, hydrogen sulfide, and particulate matter.[11] These types of emissions are typically not included in proposals to limit GHG emissions.

The sector also emits CO_2 and other gases through its on-farm energy use, for example, through the use of tractors and other farm machinery. These emissions are generally aggregated along with other transportation and industrial emissions in the "energy" sources, where they constitute a very small share of the overall total. Therefore, these emissions are not included in reported estimates for the U.S. agriculture sector.

Total GHG Emissions

In 2005, GHG emissions from U.S. agricultural activities totaled nearly 540 MMTCO_2-Eq., expressed in terms of CO_2-equivalent units, and accounted for about 7% of the total GHG emissions in the United States (**Table 1**).[12] Although the agriculture sector is a leading economic sector contributing to national GHG emissions, its share of total emissions is a distant second compared to that for the energy sector. Fossil fuel combustion is the leading source of GHG emissions in the United States (about 80%), with the energy sector generating 85% of annual emissions across all sectors.[13]

Recent trends in GHG emissions associated with the U.S. agriculture sector suggest emissions reductions in recent years. In 2005, emissions from agricultural activities were lower compared to estimates for 2000 and the most recent five-year average. However, emissions in 2005 were higher compared to reported emissions for 1990 and 1995 (**Table 1**).

Uncertainty Estimating Emissions

EPA's estimates are based on annual USDA data on crop production, livestock inventories, and information on conservation and land management practices in the agriculture sector. Actual emissions will depend on site-specific factors, including location, climate, soil type, type of crop or vegetation, planting area, fertilizer and chemical application, tillage practices, crop rotations and cover crops, livestock type and average weight, feed mix and amount consumed, waste management practices (e.g., lagoon, slurry, pit, and drylot systems), and overall farm management. Emissions may vary year to year depending on actual growing conditions. The EPA-reported data reflect the most recent data and historical updates, and reflect underlying methodological changes, in keeping with Intergovernmental Panel on Climate Change (IPCC) guidelines.[14] More detailed information is in EPA's 2007 *Inventory*.

Other Estimated Emissions

EPA's reported emissions for the U.S. agriculture sector are based on agricultural production only and do not include emissions associated with on-farm energy use and forestry activities,[15] or emissions associated with food processing or distribution. Although EPA's GHG estimates for the U.S. agriculture sector do not include CO_2 emissions from on-farm energy use, estimates of these CO_2 emissions constitute a small share of overall GHG emissions. During the last few years, EPA's estimates of CO_2 emissions from on-farm fossil fuel and mobile combustion averaged about 50 MMTCO_2-Eq. per year[16] (**Table 1**). These emissions are generally aggregated with emissions for the transportation and industrial sectors. Even if these emissions were included with other attributed GHG emissions for the agriculture sector, this would not substantially raise agriculture's overall share of total GHG emissions.

Table 1. GHG Emissions and Carbon Sinks, Agricultural Activities, 1990-2005 (CO_2-Equivalent)

Source	1990	1995	2000	2005	Avg. 2001-2005
	million metric tons CO2 equivalent (MMTCO2-Eq)				
U.S. Agricultural Activities					
GHG Emissions (CH_4 and N_2O)					
Agriculture Soil Management[a]	366.9	353.4	376.8	365.1	370.9
Enteric Fermentation[b]	115.7	120.6	113.5	112.1	115.0
Manure management	39.5	44.1	48.3	50.8	45.6
Rice Cultivation	7.1	7.6	7.5	6.9	7.4
Agricultural Residue Burning	1.1	1.1	1.3	1.4	1.2
Subtotal	**530.3**	**526.8**	**547.4**	**536.3**	**540.1**
Carbon Sinks					
Agricultural Soils	(33.9)	(30.1)	(29.3)	(32.4)	(31.7)
Other	na	na	na	na	na
Subtotal	(33.9)	(30.1)	(29.3)	(32.4)	(31.7)
Net Emissions, Agriculture	**496.4**	**496.7**	**518.1**	**503.9**	**508.4**
Attributable CO_2 emissions:[c] Fossil fuel/mobile combustion	**46.8**	**57.3**	**50.9**	**45.5**	**52.6**
%All Emissions, Agriculture[d]	**8.5%**	**8.0%**	**7.7%**	**7.4%**	**8.0%**
%Total Sinks, Agriculture	**4.8%**	**3.6%**	**3.9%**	**3.9%**	**4.0%**
%Total Emissions, Forestry	0.2%	0.2%	0.2%	0.3%	0.3%
%Total Sinks, Forestry[e]	94.3%	92.0%	94.8%	94.7%	95.0%
Total GHG Emissions, All Sectors	6,242.0	6,571.0	7,147.2	7,260.4	6,787.1
Total Carbon Sinks, All Sectors	(712.8)	(828.8)	(756.7)	(828.5)	(801.0)
Net Emissions, All Sectors	5,529.2	5,742.2	6,390.5	6,431.9	5,986.1

Source: EPA, *Inventory of U.S. Greenhouse Gas Emissions and Sinks: 1990-2005*, April 2007, [http://epa.gov/climatechange/emissions/usinventoryreport.html]. Table ES-2, Table 2-13, Table 6-1, Table 7-1, and Table 7-3. EPA data are reported in teragrams (Tg.), which are equivalent to one million metric tons each.

a. N_2O emissions from soil management and nutrient/chemical applications on croplands.

b. CH_4 emissions from ruminant livestock.

c. Emissions from fossil fuel/mobile combustion associated with energy use in the U.S. agriculture sector (excluded from EPA's reported GHG emissions for agricultural activities).

d. Does not include attributable CO_2 emissions from fossil fuel/mobile combustion.

e. Change in forest stocks and carbon uptake from urban trees and landfilled yard trimmings.

Sources of GHG Emissions

EPA's *Inventory* estimates of CH_4 and N_2O emissions from agricultural activities are measured across five categories.

- **Agriculture soil management:** Nitrous oxide emissions from farmland soils are associated with cropping practices that disturb soils and increase oxidation, which can release emissions into the atmosphere. The types of practices that contribute to emissions releases are fertilization; irrigation; drainage; cultivation/tillage; shifts in land use; application and/or deposition of livestock manure and other organic materials on cropland, pastures, and rangelands; production of nitrogen-fixing crops and forages; retention of crop residues; and cultivation of soils with high organic content.

- **Enteric fermentation:** Methane emissions from livestock operations occur as part of the normal digestive process in ruminant animals[17] and are produced by rumen fermentation in metabolism and digestion. The extent of such emissions is often associated with the nutritional content and efficiencyof feedutilizedby the animal.[18] Higher feed effectiveness is associated with lower emissions.

- **Manure management:** Methane and nitrous oxide emissions associated with manure management occur when livestock or poultry manure is stored or treated in systems that promote anaerobic decomposition, such as lagoons, ponds, tanks, or pits.

- **Rice cultivation:** Methane emissions from rice fields occur when fields are flooded and aerobic decomposition of organic material gradually depletes the oxygen in the soil and floodwater, causing anaerobic conditions to develop in the soil, which releases methane.

- **Agricultural residue burning:** Methane and nitrous oxide emissions are released by burning residues or biomass.[19]

The share of GHG emissions for each of these categories is as follows: agriculture soil management (68% of emissions), enteric fermentation (21%), manure management (10%), rice cultivation (1%), and field burning of agricultural residues (less than 1%). Approximately 70% of agricultural emissions are associated with the crop sector and about 30% with the livestock sector (**Figure 1**).[20]

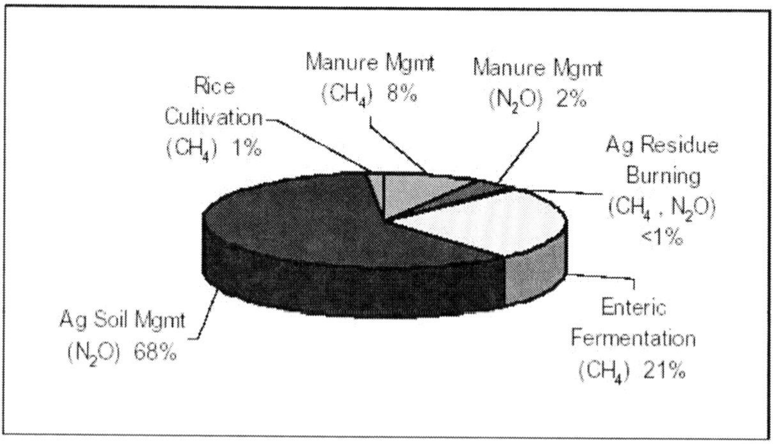

Source: EPA, 2007 *Inventory* report, April 2007, at [http://epa.gov/climatechange/emissions/usinventoryreport.html].

Figure 1. Agricultural GHG Emissions, Average 2001-2005

Potential for Additional Reductions

There is potential to lower carbon, methane, and nitrous oxide emissions from U.S. agricultural facilities at both crop and livestock operations through further adoption of certain conservation and land management practices. In most cases, such practices may both reduce emissions and sequester carbon in agricultural soils.

Improved Soil Management

Options to reduce nitrous oxide emissions associated with crop production include improved soil management, more efficient fertilization, and implementing soil erosion controls and conservation practices. In the past 100 years, intensive agriculture has caused a soil carbon loss of 30%-50%, mostly through traditional tillage practices.[21] In contrast, conservation tillage practices preserve soil carbon by maintaining a ground cover after planting and by reducing soil disturbance compared with traditional cultivation, thereby reducing soil loss and energy use while maintaining crop yields and quality. Practices include no-till and minimum, mulch, and ridge tillage. Such tillage practices reduce soil disturbance, which reduces oxidation and the release of carbon into the atmosphere. Therefore, conservation tillage practices reduce emissions from cultivation and also enhance carbon sequestration in soils (discussed later in this report). Nearly 40% of U.S. planted areas are under some type of conservation tillage practices.[22]

Improved Manure and Feed Management

Methane emissions associated with livestock production can be reduced through improved manure and feed management. Improved manure management is mostly associated with installing certain manure management systems and technologies that trap emissions, such as an anaerobic digester[23] or lagoon covers. Installing such systems generates other principal environmental benefits. Installing an anaerobic digester to capture emissions from livestock operations, for example, would also trap other types of air emissions, including air pollutants such as ammonia, volatile organic compounds, hydrogen sulfide, nitrogen oxides, and particulate matter that are regulated under the federal Clean Air Act. Other benefits include improved water quality through reduced nutrient runoff from farmlands, which may be regulated under the federal Clean Water Act.[24] Many manure management systems also control flies, produce energy, increase the fertilizer value of any remaining biosolids, and destroy pathogens and weed seeds.[25]

Manure management systems, however, can be costly and difficult to maintain, given the typically high start-up costs and high annual operating costs. For example, the initial capital cost of an anaerobic digester with energy recovery is between $0.5 million and $1 million at a large-sized dairy operation, and annual operating costs are about $36,000. Initial capital costs for a digester at a larger hog operation is about $250,000, with similar operating costs.[26] Upfront capital costs tend to be high because of site-specific conditions at an individual facility, requiring technical and engineering expertise. Costs will vary depending on site-specific conditions but may also vary by production region. Costs may be higher in areas with colder temperatures, where some types of digesters may not be appropriate or may require an additional heat source, insulation, or energy requirements to maintain constant, elevated temperatures.[27] Energy requirements to keep a digester heated are likely be lower in warmer climates.

Incentives are available to assist crop and livestock producers in implementing practices and installing systems that may reduce GHG emissions. Such incentives include cost-sharing and also low-interest financing, loan guarantees, and grants, as well as technical assistance with implementation. Funding for anaerobic digesters at U.S. livestock operations occurs under various programs under the 2002 farm bill.[28] Despite the availability of federal and/or state-level cost-sharing and technical assistance, adoption of such systems remains low throughout the United States. There are currently about 100 digester systems in operation or planned at commercial dairy and hog farms, accounting for about 1% of all operations nationwide (Figure 2).[29]

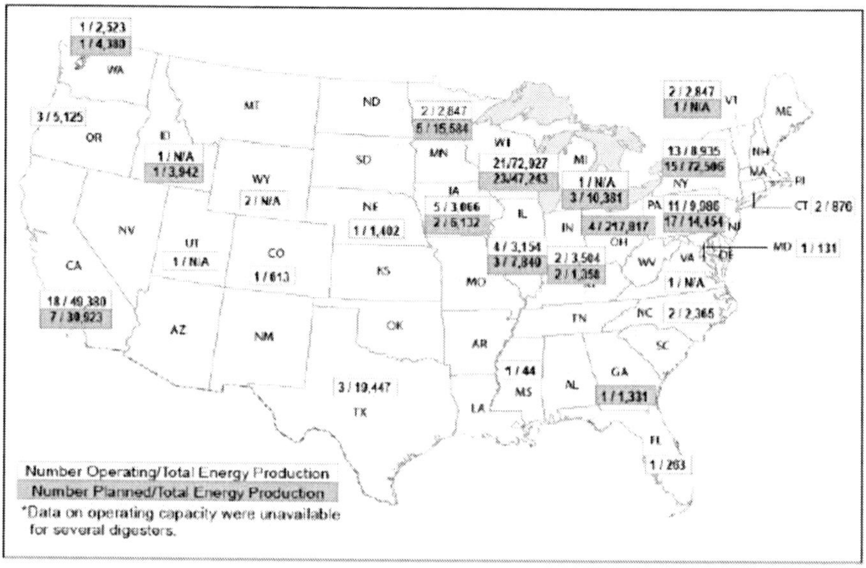

Source: Adapted by CRS, Map Resources (7/2007) from data reported by USEPA, *AgStar Digest*, Winter 2006.

Figure 2. National Distribution of Anaerobic Digester Energy Production, Operating and Planned

Improved feed strategies may also lower methane emissions at livestock operations. Such strategies may involve adding supplements and nutrients to animal diets, substituting forage crops for purchased feed grains, or instituting multi-phase feeding to improve digestive efficiency. Other options involve engineering genetic improvements in animals.[30] Purchasing feed supplements and more intensely managing animal nutrition and feeding practices may add additional costs and management requirements at the farm level.

Agricultural Carbon Sinks

Carbon Loss and Uptake

Agriculture can sequester carbon, which may offset GHG emissions by capturing and storing carbon in agricultural soils. On agricultural lands, carbon can enter the soil through roots, litter, harvest residues, and animal manure,and maybestored primarily as soil organic matter (SOM; see Figure 3).[31] Soils can hold carbon both underground in the root structure and near the soil surface and in plant biomass. Loss of soil carbon may occur with shifts in land use, with conventional cultivation (which may increase oxidation), and through soil erosion. Carbon sequestration in agricultural soils can be an important component of a climate change mitigation strategy, since the capture and storage of carbon may limit the release of carbon from the soil to the atmosphere.

Voluntary land retirement programs and programs that convert or restore grasslands and wetlands promote carbon capture and storage in agricultural soils. Related practices include afforestation (including the conversion of pastureland and cropland), reforestation, and agro-forestry practices. Conservation practices that raise biomass retention in soils and/or reduce soil disturbance, such as conservation tillage and/or installing windbreaks and buffers, also promote sequestration. More detailed information is provided in the following section, "Mitigation Strategies in the Agriculture Sector."

Total Carbon Sequestration

In 2005, carbon sequestration byagricultural soils was estimated at about 30 MMTCO_2-Eq.[32] Compared to estimates for the most recent five-year average, as well as estimates for 1995 and 2000, recent data show possible gains in carbon uptake and storage in recent years (**Table 1**).

The agriculture and forestry sectors are a small part of the overall carbon sequestration debate. Carbon sequestration by these sectors is usually referred to as indirect or biological sequestration.[33] Biological sequestration is considered to have less potential for carbon sequestration than direct sequestration, also referred to as carbon capture and storage, and is typically associated with oil and gas production.

Estimated Emission Offsets

Carbon sequestration in the U.S. agriculture sector currently offsets only about 5% of the carbon-equivalent of reported GHG emissions generated by the agriculture sector each year. Thus the sector remains a net source of GHG emissions. Compared to total national GHG emissions, the agriculture sector offsets well under 1% of emissions annually. It should be noted that these estimates do not include estimates for the forestry sector, or sequestration activities on forested lands or open areas that may be affiliated with the agriculture sector. Forests and trees account for a majority (about 95%) of all estimated carbon uptake in the United States, mostly through forest restoration and tree-planting.[34] Carbon uptake in soils on U.S. agricultural lands accounts for the bulk of the remainder.

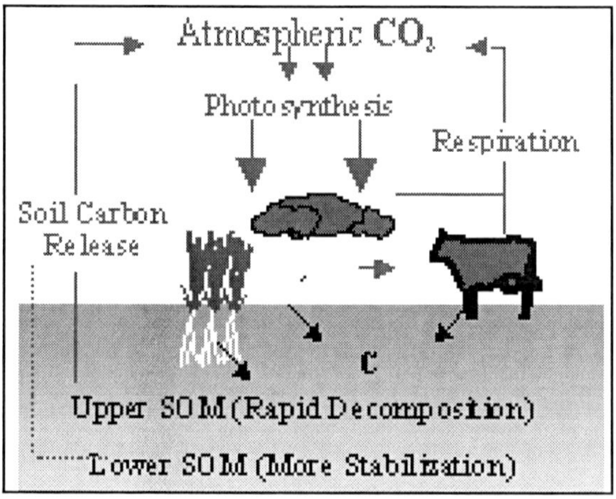

Source: USGS, "Carbon Sequestration in Soils." SOM = Soil organic matter

Figure 3. Carbon Sequestration in Agricultural Soils

Uncertainty Estimating Carbon Sinks

EPA's *Inventory* estimates of carbon uptake in agricultural soils are based on annual data and information on cropland conversion to permanent pastures and grasslands, reduced summer fallow areas in semi-dry arcas, increased conservation tillage, and increased organic fertilizer use (e.g, manure) on farmlands, as well as information on adoption rates and use of certain conservation and land management practices.

However, actual carbon uptake in agricultural soils depends on several site-specific factors, including location, climate, land history, soil type, type of crop or vegetation, planting area, tillage practices, crop rotations and cover crops, and farm management in implementing certain conservation and land management practices. Estimates of the amount of carbon sequestered may vary depending on the amount of site-specific information included in the estimate, as well as on the accounting procedures and methodology used to make such calculations.

In general, the effectiveness of adopting conservation and land management practices will depend on the type of practice, how well the practice is implemented, and also on the length of time a practice is undertaken. For example, time is needed for a certain conservation practice to take hold and for benefits to accrue, such as buildup of carbon in soils from implementing conservation tillage or other soil management techniques, and growing time for cover crops or vegetative buffers. The overall length of time the practice remains in place is critical, especially regarding the sequestration benefits that accrue over the time period in which land is retired. In addition, not all conservation and land management practices are equally effective or appropriate in all types of physical settings. For example, the use and effectiveness of conservation tillage practices will vary depending on soil type and moisture regime, which may discourage some farmers from adopting or continuing this practice in some areas.

The potential impermanence of conservation and land management practices raises concerns about the effectiveness and limited storage value of the types of conservation practices that sequester carbon, given that the amount of carbon stored depends on the willingness of landowners to adopt or continue to implement a particular voluntary conservation practice. There are also concerns that the addition of other conservation practices may not significantly enhance the sequestration potential of practices that might already be in place.[35] This raises questions about the cost-effectiveness of sequestering carbon on farmlands relative to other climate change mitigation strategies in other industry sectors. Finally, implementing conservation practices and installing new technologies may be contingent on continued cost-sharing and other financial incentives contained in the current farm bill; programs funded through this legislation help offset the cost to farmers for these practices and technologies, which some farmers may not be willing to do otherwise.

Potential for Additional Uptake

USDA reports that the potential for carbon uptake in agricultural soils is much greater than current rates. USDA forecasts that the amount of carbon sequestered on U.S. agricultural lands will more than double from current levels by 2012, adding roughly an additional 40 MMTCO_2-Eq. of sequestered carbon attributable to the sector.[36] This additional uptake is expected through improved soil management (roughly 60%), improved manure and nutrient management (about 30%), and additional land-retirement sign-ups (about 10%).

Other longer-term estimates from USDA report that the potential for net increases in carbon sequestration in the agriculture sector could range from 40 to 590 MMTCO_2-Eq. per year, or roughly 2-20 times current levels.[37] Afforestation, or the creation of forested areas mostly through conversion of pastureland and cropland, reflects the majority of the estimated uptake potential, with agricultural soil carbon sequestration accounting for a smaller share at the high end of this estimated range. Comparable estimates reported by EPA forecast a higher sequestration potential for the U.S. agriculture sector, ranging from 160 to 990 MMTCO_2-Eq. per year.[38] EPA also reports additional sequestration potential from livestock manure management, biofuels substitution, and forest land management. Estimates from various studies may differ depending on the extent that estimates may include sequestration activities in the forestry sector. Combined, the potential carbon uptake from both the agriculture and forestry sectors is estimated from 800 to 1,200 MMTCO_2-Eq. per year.[39]

An additional carbon uptake potential of 590 to 990 MMTCO_2-Eq. per year would more than offset the agriculture sector's annual GHG emissions, or offset 8% to 14% of total current national emissions from all sources. Currently, carbon uptake in agricultural soils sequesters under 1% of total national GHG emissions annually (**Table 1**). An estimated 11% of all GHG emissions are currently sequestered annually, with the bulk sequestered through growth in forest stocks.

Per-Unit Value Estimates

Compared to other mitigation options in other sectors, USDA reports that U.S. agriculture can provide low-cost opportunities to sequester additional carbon in soils and biomass. The estimated per-unit value (or cost) of carbon removed or sequestered, expressed on a dollar per metric ton (mt) of carbon basis, will vary depending on the type of practice. Actual per-unit values and the cost-effectiveness of different practices may vary considerably from site to site.

USDA's estimate of an additional carbon uptake potential of 40 to 590 MMTCO_2-Eq. per year is associated with a range of costs from about \$3/mt to \$35/mt of permanently sequestered carbon dioxide (**Table 2**).[40] The low end of this range reflects the sequestration potential associated with cropland management practices; higher-end values are associated with land retirement and conversion, and a longer sequestration tenure. USDA's report also notes that if producers discontinue the land and cropland management practices at the end of a typical contract period, the carbon sequestered may only be worth a small share of its overall program costs, because most of the carbon will be released when these practices are terminated, which may lower the cost-effectiveness of such programs. EPA's forecast of an additional sequestration potential for the agriculture sector of 160 to 990 MMTCO_2Eq. per year are estimated across a range of $5/mt-$30/mt of sequestered carbon dioxide.[41] The low end of this range is associated with sequestration in agricultural soils and with soil management practices; high-end values are associated with afforestation, or converting open land into a forest by planting trees or their seeds.

Table 2. Carbon Sequestration Potential in the U.S. Agriculture Sector, Alternative Scenarios and Payment Levels (dollars per million metric ton of sequestered CO_2)

Source	$3-5 range	$14-15 range	$30-34 range
	(million mt of sequestered CO2)		
USDA Estimate			
Afforestation	0 - 31	105 - 264	224 – 489
Agricultural soil carbon sequestration	0.4 - 4	3 - 30	13 - 95
Total	**0.4 - 35**	**108 - 295**	**237 - 587**
EPA Estimate			
Afforestation	12	228	806
Agricultural soil carbon sequestration	149	204	187
Total	**161**	**432**	**994**

Sources: EPA, *Greenhouse Gas Mitigation Potential in U.S. Forestry and Agriculture*, Nov. 2005, Table 4-10, at [http://www.epa.gov/sequestration/greenhouse_gas.html]. Compares USDA estimates (*Economics of Sequestering Carbon in the U.S. Agricultural Sector*, Apr. 2004) with EPA estimates.

Enhancing Carbon Sinks

There is potential to increase the amount of carbon captured and stored in U.S. agricultural lands by adopting certain conservation and land management practices. In most cases, such practices may both sequester carbon in farmland soils and reduce emissions from the source. **Table 3** shows estimated representative carbon sequestration rates for agricultural practices.

Improved Soil and Land Management

The main carbon sinks in the agriculture sector are cropland conversion and soil management, including improved manure application.[42] More than half of all carbon sequestered on U.S. agricultural lands is through voluntary land retirement programs and programs that convert or restore land (e.g., conversion to open land or grasslands, conversion to cropland, restoration of grasslands or wetlands, etc.). Undisturbed open lands, grasslands and wetlands can hold carbon in the soil both underground in the root structure and above ground in plant biomass. The amount of carbon sequestered will vary by the type of land management system. Afforestation and cropland conversion have the greatest potential to store the most carbon per acre annually, compared with other types of systems, such as tree plantings and wetlands conversion, or storage in croplands.[43]

Table 3. Representative Carbon Sequestration Rates

Type of land Management System	Sequestration Rate ($mt\ CO_2$/acre/year)
Afforestation	2.2 - 9.5
Reforestation	1.1 - 7.7
Reduced tillage (e.g., no-till, reduced-till)	0.6 - 1.1
Change in grassland management	0.07 - 1.9
Cropland conversion to grassland	0.9 - 1.9
Riparian buffers (nonforest)	0.4 - 1.0
Biofuel substitution for fossil fules	4.8 - 5.5

Source: Compiled by EPA, *Greenhouse Gas Mitigation Potential in U.S. Forestry and Agriculture*, Table 2-1, Nov. 2005, at [http://www.epa.gov/sequestration/greenhouse_gas.html]. Saturation rates and duration periods apply. EPA's report provides a list of the original source citations.

Conservation tillage is another major source of sequestration on farmlands, accounting for about 40% of the carbon sequestered by the U.S. agriculture sector.[44] Improved tillage practices improve biomass retention in soils and reduce soil disturbance, thereby decreasing oxidation. The amount of carbon sequestered will varybythe type of tillage system: reduced tillagestores between 0.6-1.1 mt of carbon dioxide per acre annually (**Table 3**). Among conservation tillage practices, no-till stores about 30% more than the amount of carbon stored by reduced tillage but more than five times that stored on intensive tilled croplands. (Conservation tillage practices are explained in the section on "Potential for Additional Reductions").

Improved Manure and Feed Management

Mitigation strategies at U.S. livestock operations are not commonly associated with carbon uptake and are not included in EPA's carbon sink estimates. However, installing manure management systems, such as an anaerobic digester, captures and/or destroys methane emissions from livestock operations and may be regarded as avoided emissions or as a form of direct sequestration capturingemissions at the source. Asa result, some carbon offset programs are beginning to promote manure management systems as a means to capture and store methane at dairy operations, which may also be sold as carbon offset credits and as a renewable energy source.[45] Given that there are currently few anaerobic digesters in operation, estimates of the actual or potential uptake may be difficult to estimate. (Manure management systems are further explained in the section on "Potential for Additional Reductions.")

MITIGATION STRATEGIES IN THE AGRICULTURE SECTOR

Existing conservation and farmland management programs administered at both the federaland state levels often encourage the types of agriculturalpractices thatcan reduce GHG emissions and/or sequester carbon. These include conservation, forestry, energy, and research programs within existing farm legislation. These programs were initiated predominantly for other production or environmental purposes, and few specifically address climatechangeconcerns in the agriculture and forestry sectors. However, some USDA and state-level programs have started to place additional attention on the potential for emissions reduction and carbon storage under certain existing programs.

Agricultural conservation and other farmland practices broadly include land management, vegetation, and structures that can also reduce GHG emissions and/or sequester carbon, such as:

- land retirement, conversion, and restoration (e.g., conversion to grasslands, restoration of grasslands or wetlands, etc.);
- soil conservation practices, including conservation tillage (e.g., reduced/medium- till, no/strip-till, ridge-till);
- soil management and soil erosion controls;
- precision agriculture practices and recognized agricultural best management practices;
- efficient fertilizer/nutrient (incl. manure) and chemical application;

- crop rotations;
- cover cropping;
- manure management (e.g., improve manure storage and technologies using anaerobic digestion and methane recovery);
- feed management (e.g., improve feed efficiency, dietary supplements);
- rotational grazing and improved forage/grazing management;
- vegetative and riparian buffers, and setbacks;
- windbreaks for crops and livestock;
- bioenergy and biofuels substitution and renewable energy use (e.g., replacing use of fossil fuels); and
- energy efficiency and energy conservation on-farm.

In general, conservation programs administered by USDA and state agencies encourage farmers to implement certain farming practices and often provide financial incentives and technical assistance to support adoption. Participation in these programs is voluntary, and farmers may choose to discontinue participating in these programs. The effectiveness of these practices depends on the type of practice, how well the practice is implemented, and also on the length of time a practice is undertaken. These programs are generally designed to address site-specific improvements based on a conservation plan developed with the assistance of USDA or state extension technical and field staff that considers the goals and land resource base for an individual farmer or landowner. Such a conservation plan is typically a necessary precursor to participating in USDA's conservation programs.

Federal Programs

Conservation Programs

Conservation programs administered by USDA are designed to take land out of production and to improve land management practices on land in production, commonly referred to as "working lands" (**Table 4**). These programs are provided for in Title II (Conservation) of the 2008 farm bill.

- **Land retirement/easement programs.** Programs focused on land management, including programs that retire farmland from crop production and convert it back into forests, grasslands, or wetlands, including rental payments and cost-sharing to establish longer term conservation coverage. Major programs include the Conservation Reserve Program (CRP), the Wetlands Reserve Program (WRP), the Grasslands Reserve Program (GRP), the Farmland Protection Program (FPP), among other programs.

- **Working lands programs.** Programs focused on improved land management and farm production practices, such as changing cropping systems or tillage management practices, are supported by cost-sharing and incentive payments, as well as technical assistance. Major programs include the Environmental Quality Incentives Program (EQIP), the Conservation Stewardship Program (CSP), the Agricultural Management Assistance (AMA) program, and the Wildlife Habitat Incentives Program (WHIP).

Table 4. Conservation and Land Management Practices

USDA Program	Conservation Practice and Land Management	General Benefits	Benefits for Climate Change
EQIP, CSP, AMA	Conservation tillage and reduced field pass intensity	Improves soil/water/air quality. Reduces soil erosion/fuel use.	Sequestration, emission reduction
	Crop diversity through crop rotations and cover cropping	Reduces erosion/water needs. Improves soil/water quality.	Sequestration
	Efficient nutrient (nitrogen) management, fertilizer application	Improves water quality. Saves expenses, time, and labor.	Sequestration, emission reduction
	Improved soil management and soil erosion controls	Improves soil/water/air quality.	Sequestration, emission reduction
EQIP CSP AMA Other[a]	Manure management (e.g., storage/ containment, anaerobic digestion and methane recovery)	Improves soil/water/air quality. On-farm fuel cost-savings. Alternative income source. Nutrients for crops.	Emission reduction
EQIP CSP AMA	Feed management (e.g., raise feed efficiency, dietary supplements)	Improves water/air quality. More efficient use of feed.	Emission reduction
	Rangeland management (e.g., rotational grazing, improved forage)	Reduces water requirements. Helps withstand drought. Raises	Sequestration, emission reduction

		grassland productivity.	
EQIP CSP AMA WHIP	Windbreaks for crops and livestock, vegeta-tive/riparian buffers, grassed waterways, setbacks, etc.	Improves crop/livestock protection and wildlife habitat. Alter-native income source (e.g., hunting fees).	Sequestration, emission reduction
EQIP CSP AMA	Agroforestry / silvopasture with rotational grazing and improved forage	Provides income from grazing and wood products.	Sequestration, emission reduction
CRP WRP GRP FPP	Land management, including retirement, conversion, restoration (cropland, grasslands, wetlands, open space)	Improves soil/ water/air quality.	Sequestration
EQIP CSP AMA Other[a]	Energy efficiency/ conservation	Improves soil/water/air quality. Cost-savings.	Emission reduction
	Biofuel substitution and renewable energy use	Improves soil/water/air quality. On-farm fuel cost-savings. Alternative income source.	Emission reduction

Source: Compiled by CRS staff from available USDA and EPA information. Listed programs: Conservation Reserve Program (CRP), Wetlands Reserve Program (WRP), GrasslandsReserve Program (GRP), Farmland Protection Program (FPP), Environmental Quality Incentives Program (EQIP), Conservation Stewardship Program (CSP), Agricultural Management Assistance (AMA), Wildlife Habitat Incentives Program (WHIP).

a. Renewable energy projects receive additional program funding in farm bill under Title IX (Energy) and Title VI (Rural Development), as well as other federal and state program.

Prior to the 2008 farm bill, few USDA conservation programs were specifically intended to address climate change concerns in the agriculture sector. One exception is USDA's Conservation Innovation Grants program, a subprogram under EQIP that provides for competitive awards, and is intended to accelerate technology transfer and adoption of innovative conservation technologies, mostly through pilot projects and field trials. Past grants have supported development of approaches to reduce ammonia emissions from poultry litter, promote conservation tillage and solar energy technologies, and develop private carbon sequestration trading credits.[46]

However, USDA has considered expanding three of its existing conservation programs — CRP, EQIP, and CSP — in ways that could further encourage emission reductions and carbon sequestration.[47] For example, USDA notes that many of the practices encouraged under EQIP and CSP reduce net emissions. For EQIP, USDA is providing additional guidance to technical staff to make GHG a priority resource concern as part of its ranking system and scoring criteria for participation by, for example, giving greater weight to projects that promote anaerobic digestion, nutrient management plans, and other types of cropland practices, such as installing shelter belts and windbreaks, encouraging conservation tillage, and providing resources for biomass energy projects. Under CRP, USDA has issued a new rule that explicitly allows the private sale of carbon credits for land enrolled in the program. It also modified how it scores and ranks offers to enroll land in CRP in order to place greater weight on installing vegetative covers that sequester carbon. USDA also has announced a program under CRP's continuous enrollment provision to plant up to 500,000 acres of bottomland hardwoods, which are among the most productive U.S. lands for sequestering carbon.

Not including funding increases authorized under the 2008 farm bill, actual total funding in FY2005 for USDA's conservation programs totaled $5.6 billion. Voluntary land retirement programs and programs that convert or restore land account for roughly 37% annually of all USDA conservation spending (Figure 4). Programs that provide cost-sharing and technical assistance to farmers to implement certain practices, such as EQIP, CSP, and AMA, provide another 21% annually. USDA's conservation technical assistance and extension services account for about one-fourth of all funding. Other federal funding through other programs also generally promotes natural resource protection on U.S. farms. Generally, the decision on how and where this fundingis ultimatelyused is made at the individual state level.

The new 2008 farm bill (P.L. 110-246, the Food, Conservation, and Energy Act of 2008) expands several existing conservation programs that contribute to increased carbon storage in soil and plants, reduced agriculture-based emissions associated with climate change, lowered energy consumption by farming operations, and increased production of renewable fuels and feedstocks, among other provisions.

The 2008 farm bill increases funding for both EQIP and CSP, and provides for expanded eligibility to include management practices on private forest lands and other natural resource areas. It also provides funding for the Conservation Innovation Grants program to address air quality concerns from agriculture operations, including greenhouse gas emissions. The farm bill also makes changes to USDA's land retirement programs. Changes to CRP will encourage the establishment of native vegetation cover on lands set aside or retired from agricultural production, and promote tree planting and management to improve habitat and encourage healthy forest growth and carbon uptake. Changes to FPP include expanded eligibility for forest lands, and changes to GRP include expanded grasslands enrollment and emphasis on long-term and permanent easement.

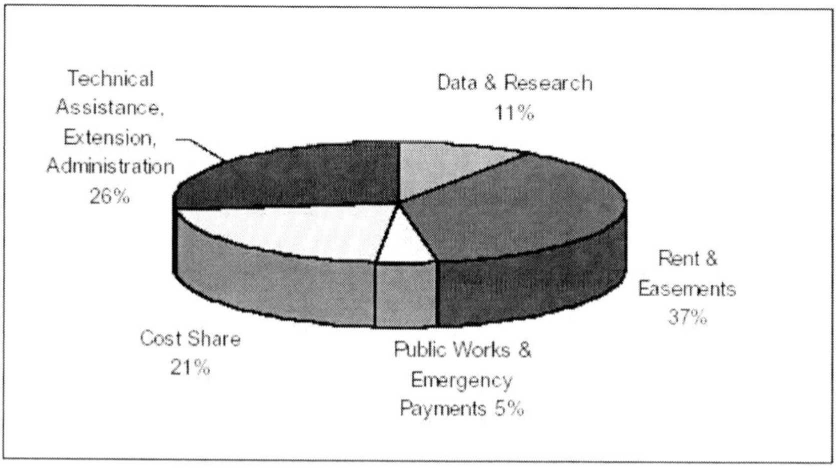

Source: USDA, Office of Budget and Planning.
Note: FY2005 total spending = $5.6 billion.

Figure 4. USDA Conservation Spending, FY2005

The farm bill also creates a new conservation provision to facilitate the participation of farmers and ranchers in emerging carbon and emissions trading markets by directing USDA to establish guidelines for standards, accounting procedures, reporting protocols, and verification processes for carbon storage and other types of environmental services markets. (This new provision is described in further detail in the section on "Farm Bill Legislation.")

Other Farm Programs

Aside from USDA's conservation programs, there are other farm bill programs that encouragethetypes of agricultural practices that can reduce GHG emissions and/or sequester carbon. These include programs in the farm bill's forestry, energy, and research titles.[48]

Renewable energy projects receive additional program funding across three farm bill titles: Title II (Conservation), Title IX (Energy), and Title VII (Research). In addition to cost-sharing provided under USDA's conservation programs, one energy title provision in the 2008 farm bill is the so-called Rural Energy for America Program (Section 9007). This program provides mandatory funding for grants for energy audits, renewable energy development, and financial assistance to promote energy efficiency and renewable energy development for farmers and rural small businesses.[49] In the past this program has provided funding to support construction of anaerobic digesters in the livestock sector.[50] Other renewable energy funding is also available through other federal programs.[51] The 2008 farm bill also created the Biomass Crop Assistance Program to assist in the development of renewable energy feedstocks, including cellulosic ethanol, and to provide incentives for producers to harvest, store, and transport biomass. The farm bill's Title VII (Research) also provides for research on renewable fuels, feedstocks, and energy efficiency and for competitive grants for on-farm research and extension projects.

Forestry programs, administered by USDA's Forest Service, are provided for in Title VIII (Forestry) of the farm bill. Typically, there is often little overlap between the various agriculture and forestry programs administered by USDA, and few forestry programs provide support to agricultural enterprises.[52] One program with an agroforestry component is the Healthy Forests Reserve Program, which was reauthorized in the 2008 farm bill. This program assists with restoring and enhancing forest ecosystems; however, funding for this program is usually limited to a few states. The 2008 farm bill also created new programs with possible agroforestry benefits, including (1) the Community Forest and Open Space Conservation Program, authorizing new cost-share grants for local governments, tribes, and non-profits to acquire lands threatened by conversion to non-forest uses; and (2) the Emergency Forest Restoration Program, providing for the rehabilitation of croplands, grasslands, and private non-industrial forests following natural disasters. The farm bill also expanded or created other programs to protect and restore privately owned forests, which could also contribute to retaining or increasing carbon storage capacity on forest lands.

State Programs

State-level agriculture conservation and land management programs are available to farmers in most states, and operate in much the same manner as federal conservation programs. These programs may also provide financial and technical assistance to farmers to implement certain practices, using additional state resources and in consultation with state agriculture agencies and extension staff. No single current compendium exists outlining the different types of agriculture conservation programs across all states; instead information is available through individual state government websites.[53]

Many states have cost-share programs that provide financial assistance to landowners to implement practices that benefit a state's forests, fish, and wildlife. Many of these programs provide technical assistance and up to 75% of the eligible costs of approved conservation projects to qualified landowners. Several states also provide low-interest financing to farmers and landowners to encourage conservation practicesor to implement best management practices for theagriculturesector. Many states also have buffer strip programs, which may provide rental payments to landowners who agree to create or maintain vegetative buffer strips on croplands near rivers, streams, ponds, and wetlands. Typically states that have taxing authority for conservation purposes, such as Nebraska, Missouri, and Oregon, tend to have more stable funding and staffing to support conservation improvements.

The Pew Center on Global Climate Change has identified several ongoing state programs and demonstration projects specifically intended to promote carbon storage and emissions reduction in the U.S. agriculture sector.[54] For example, several states, including Oregon, Wisconsin, Vermont, and North Carolina, are promoting methane recovery and biofuels generation from livestock waste. A program in Iowa is providing support and funding to promote switchgrass as a biomass energy crop. In Maryland, income tax credits are provided for the production and sale of electricity from certain biomass combustion. Georgia has a program that leases no-till equipment to farmers. In addition, several states, including Nebraska, Oklahoma, Wyoming, North Dakota, and Illinois, have formed advisory committees to investigate the potential for state carbon sequestration. In California, an accounting program is being developed to track possible future costs to mitigate GHG emissions in the U.S. agriculture sector.

An even greater number of state programs and initiatives are geared toward climate change mitigation strategies in sectors other than agriculture.[55] For example, many of California's programs support the state's recently enacted emission reductions legislation.[56] California's climate change statute requires state agencies to identify GHG emissions reduction strategies that can be pursued before most of the law takes effect in 2012. The state has identified several agriculture sector strategies that it plans to consider as early actions, including (1) adopting a manure digester protocol for calculating GHG mitigation; (2)

establishing collaborative research on how to reduce GHG emissions from nitrogen land application; (3) replacing stationary diesel agricultural engines with electric motors; and (4) evaluating potential measures for enclosed dairy barns, modified feed management, and manure removal strategies to reduce methane emissions at dairies.[57] These early action strategies would be in addition to funding for the state's manure digester cost-share program and other agriculture projects, including carbon sequestration projects involving rice straw utilization, energy and water conservation, biofuels support, soil management, and other types of renewable energy and manure management programs for dairies.[58]

Other Programs and Incentives

The voluntary carbon offset market allows businesses, interest groups, and individuals the opportunity to purchase carbon credits generated from projects that either prevent or reduce an amount of carbon entering the atmosphere, or that capture carbon from the atmosphere. Companies and individuals purchase carbon credits for varied reasons. For example, some may purchase credits to reduce their "carbon footprint," using credits to offset all or part of a GHG-emitting activity (e.g., air travel, corporate events, or personal automobile use); others may purchase credits to bank the reductions in anticipation of a mandatory GHG reduction program.[59] In the United States, the current offset framework operates on a voluntary basis since there is no federal requirement that GHG emissions be curtailed. Some states and/or regional GHG reduction initiatives may limit the use of carbon offsets.

Several states have programs that support the voluntarycarbon offset exchange, often involving U.S. farmers and private landowners. One program operated by the Iowa Farm Bureau involves more than 1,400 producers in 12 states (mostly Iowa, Kansas, and Nebraska, but also Illinois, Ohio, Michigan, Wisconsin, Minnesota, South Dakota, Missouri, Indiana, and Kentucky),[60] whose carbon credits may be sold on the Chicago Climate Exchange.[61] Similar types of programs also have been initiated in North Dakota (operated by the North Dakota Farmers Union), Illinois (Illinois Conservation and Climate Initiative), Indiana (Environmental Credit Corporation), and the Northwest (Upper Columbia Resource Conservation and Development Council). Another, Terrapass, has among its projects two large-scale dairyfarms that use anaerobic digesters and methane capture for energy production.[62]

Farmer participation in voluntary carbon credit trading programs has been growing rapidly. As of early 2008, participation involved an estimated 4,000 farmers across 25-30 states covering more than 4 million acres.[63] The two largest programs providing for farm-based offsets are programsoperated by the Iowa Farm Bureau and the North Dakota Farmers Union.[64] Farm-based offset programs generally cover some or all aspects of the following types of carbon capture and storage activities: sustainable agriculture practices (such as conservation tillage, grass seedlings); planting of unharvested grasslands; tree plantings; methane capture/biogas production with manure digesters; wind, solar, or other renewable energy use; controlled grasslands or pasture management; and forest restoration. Farmer participation in such programs may help offset farm costs to install emissions controls and/or practices that sequester carbon by providing a means for them to earn and sell carbon credits.

RECENT CONGRESSIONAL ACTION

Congress is considering a range of climate change policy options, including mandatory GHG emission reduction programs. The current legislative proposals generally would not require emission reductions in the agriculture and forestry sectors. However, some of the GHG proposals would allow for regulated entities (e.g., power plants) to purchase carbon offsets, including those generated in the agriculture and forestry sectors. These and related bills and issues are currently being debated in Congress. Some of these proposals dovetail with provisions that were enacted as part of the 2008 farm bill, including a provision that directs USDA to develop guidelines and standards for quantifying carbon storage by the agriculture and forestry sectors, among other farm bill provisions that indirectly encourage emissions reductions and carbon capture and storage.

Climate Change Legislation

During the 110th Congress, several proposals have been introduced that would either mandate or authorize a cap-and-trade program to reduce GHG emissions. A cap-and-trade program provides a market-based policy tool for reducing emissions by setting a cap or maximum emissions limit for certain industries. Sources covered by the cap can choose to reduce their own emissions, or can choose to buy emission credits that are generated from reduction made by other sources. Applying this type of market-based approach to GHG reductions and trading would be similar to the acid rain reduction program established by the

1990 Clean Air Act Amendments. For more information about these GHG legislative proposal, see CRS Report RL33846, *Greenhouse Gas Reduction: Cap-and-Trade Bills in the 110th Congress*, and CRS Report RL34067, *Climate Change Legislation in the 110th Congress*, by Brent D. Yacobucci and Jonathan L. Ramseur.

Source of Emissions Reductions

Historically, climate-related legislative initiatives have not specifically focused on emissions reductions in the agriculture sector. In part, this may reflect the general consensus, as reflected by the House Energyand Commerce Committee, that GHG "emissions from the agriculture sector generally do not lend themselves to regulation under a cap-and-trade program," given the "large number of sources with small individual emissions that would be impractical to measure."[65]

In general, the current legislative proposals do not include the agriculture sector as a covered industry, which would require farmers and landowners to reduce emissions associated with climate change. One exception is H.R. 6186 (Markey), which would require performance standards for certain sources of methane and nitrous oxide emissions, including animal feeding operations; H.R. 6186 would specifically not include crop operations and forest management systems. However, some interest groups continue to question whether certain types of agricultural operations could eventually be brought in under some proposals. For example, some bills would provide authority to EPA to determine covered entities by applying cost-effective criteria to reduction options;[66] other bills such as S. 3036 (Leiberman/Warner) would cover biogenic emissions resulting from biological processes, which some interpret as potentially including animal agriculturefacilities. Still others argue that U.S. agriculture will be affected by anticipated climate legislation in terms of generally increasing energy and production input costs that will negatively impact the farming sector.[67]

Source of Offsets and Allocations

Several of the cap-and-trade proposals do incorporate the agriculture and forestry sectors either as a source of carbon offsets[68] or as a recipient of set-aside allowances.[69] In the context of these legislative proposals, a carbon offset is a measurable avoidance, reduction, or sequestration of carbon dioxide (CO_2) or other GHG emissions, expressed in carbon-equivalent terms.[70] A set-aside allowance refers to a set percentage of available allowances under the overall emissions cap that is allocated to non-regulated entities, in this case domestic agriculture and forestry entities. Some bills also specify that the proceeds from auctioned allowances be used to promote certain objectives, which could further encourage farmland conservation and bio-energy technologies and practices, among other activities.[71]

Many of the GHG bills — S. 280 (McCain/Lieberman), S. 317 (Feinstein), S. 1168 (Alexander/Lieberman), S. 1177 (Carper), S. 1766 (Bingaman/Specter), S. 3036 (Lieberman/Warner), and H.R. 620 (Olver) — would allow for the use of carbon offsets, including agricultural activities and other land-based practices, under a cap-and-trade framework. This builds on the concept, also expressed by the House Energy and Commerce Committee, that emissions reductions and carbon sequestration by the agriculture sector may provide an appropriate source of credits or offsets within a cap-and-trade program.[72] Some bills — S. 309 (Sanders/Boxer), S. 485 (Kerry), S. 1201 (Sanders), S. 1554 (Collins/Lieberman), and H.R. 1590 (Waxman) — would not allow for offsets, but would set aside a percentage of allowances for various purposes, including biological sequestration. Participating farmers and landowners who receive these allowances for sequestration and/or emission reduction activities could sell them to facilities that could become covered by a cap-and-trade program.

For example, a Senate bill reported by the Senate Committee on Environment and Public Works (EPW), the Lieberman-Warner Climate Security Act of 2008 (S. 3036, formerly S. 2191), contains several agriculture- and forestry-based provisions. The cap-and-trade framework outlined in S. 3036 establishes a tradeable allowance system that includes a combination of auctions and free allocation of tradeable allowances. As part of this overall framework, S. 3036 includes three design mechanisms that may provide financial incentives to encourage land-based agricultural and forestry activities: carbon offsets, set-aside allowances, and auction proceeds. S. 3036 provides for a range of agriculture and forestry offset projects, including agricultural and rangeland sequestration and management practices, land use change and forestry activities, manure management and disposal, and other terrestrial offset practices identified by USDA. S. 3036 also would directly allocate 5% of the overall emissions allowances to domestic agriculture and forestry entities, and allocate a set percentage of available auction proceeds to carry out a cellulosic biomass ethanol technology deployment program. For more information on the agriculture and forestry provisions in S. 3036, see CRS Report RS22834, *Agriculture and Forestry Provisions in Climate Change Legislation (S. 3036)*, by Renée Johnson.

The inclusion of these types of provisions could provide opportunities to some farmers and landowners by allowing them to directly participate in and potentially gain a significant part of this emerging carbon market. The offset and allowance provisions could allow farmers and landowners to participate in the emergingmarket by granting them the use of allowances and credits for sequestration and/or emission reduction activities. These allowances and credits could be sold to regulated facilities (e.g., power plants) covered by a cap-and-trade program to meet their emission reduction obligations. Proceeds from the sale of these allowances, credits, and auctions could be used to further promote and support activities in these sectors that reduce, avoid, or sequester emissions.

The inclusion of provisions that allow for agriculture and forestry offset and allowances as part of a cap-and-trade scheme is generally supported by a broad-based industry coalition. This coalition consists of agricultural groups representing commodity crops, livestock and dairy, the American Farm Bureau Federation, the National Farmers Union, the American Farmland Trust, and other agriculture support and utility companies.[73] Former Senators and Majority Leaders Bob Dole and Tom Daschle are also advocating on behalf of the Bipartisan Policy Center that farmers be fully integrated into any cap-and-trade scheme.[74]

However, the inclusion of carbon offsets from the agriculture and forestry sectors within a cap-and-trade program has remained controversial since the Kyoto Protocol negotiations.[75] During those negotiations, there was marked disagreement among countries and interest groups, arguing either for or against the inclusion of offsets from the agriculture and forestry sectors.[76]

The EU's GHG emission program, the Emission Trading System (ETS), which was established in 2005, does not provide for agricultural or forestry projects and activities. Among the reasons are (1) pragmatic concerns regarding measurement and verification, given the sheer number of farmers and landowners, and (2) ideological concerns about granting too much flexibility in how emission reductions are met, which could undermine overall program goals. Among the areas of concern regarding biological sequestration offsets are those highlighted in two previous sections of this report, "Uncertainty Estimating Emissions" and "Uncertainty Estimating Carbon Sinks." In summary, primary areas of concern include:

- **Permanence/Duration,** given that land uses can change over time (e.g., forest lands to urban development, other natural events such as fires or pests);

- **Measurement/Accounting,** given that biological sequestration measurement is difficult and estimates can vary, actual emission reduction or sequestration depends on site-specific factors (e.g., location, climate, soil type, crop/vegetation, tillage practices, farm management, etc.);

- **Effectiveness,** the success of the mitigation practice will depend on the type of practice, how well implemented and managed by the farmer or landowner, and the length of time the practice is undertaken;

- **Additionality,** given that some of the activities generating offsets would have occurred anyway under a pre-existing program or practice, and thus may not go beyond business as usual (BAU);

- **Leakage,** given that reductions in one place could result in additional emissions elsewhere; and

- **Double counting,** given that some reductions may be counted by another program (e.g., attributable to other environmental goals under various farm conservation programs) or towards more than one GHG reduction target.

A more detailed discussion of some of these issues is available in CRS Report RL34436, *The Role of Offsets in a Greenhouse Gas Emissions Cap-and-Trade Program: Potential Benefits and Concerns*, by Jonathan L. Ramseur. Many of these concerns, as well as the potential market opportunities issues for farmers and landowners, were discussed at a subcommittee hearing of the Senate Agriculture Committee in May 2008.[77]

Farm Bill Legislation

To help address some of the measurement and quanitification issues surrounding agricultural and forestry carbon credits, both the House and Senate Agriculture committees included a specific provision in their respective versions of the new omnibus farm bill to address this issue. This and other related provisions were included in the enacted 2008 farm bill (P.L. 110-246, the Food, Conservation, and Energy Act of 2008). Specifically, the enacted bill contains a new conservation provision that seeks to facilitate the participation of farmers and landowners in environmental services markets, including carbon storage. The bill also expands existing voluntary conservation and other farm bill programs, providing incentives that could accelerate opportunities for agriculture and forestry to reduce emissions associated with climate change, adopt energy efficiency measures, and produce renewable energy feedstocks.

In particular, the new environmental services market provision seeks to "establish technical guidelines that outline science-based methods to measure the environmental services benefits from conservation and land management activities in order to facilitate the participation of farmers, ranchers, and forest landowners in emerging environmental services markets." The intended purpose of these technical guidelines is to develop (1) a procedure to measure environmental services benefits; (2) a protocol to report environmental services benefits; and (3) a registry to collect, record and maintain the benefits measured. The provision also requires that USDA provide guidelines for establishing a verification process as part of the protocol for reporting environmental services, but it allows USDA to consider the role of third parties in conducting independent verification. In carrying out this directive, USDA is directed to work in consultation with other federal and state government agencies, non-governmental interests, and other interested persons as determined by USDA. However, the enacted bill does not specifically address funding for this provision. Nevertheless, the inclusion of this provision in the farm bill will expand the scope of existing farm and forestry conservation programs in ways that will more broadly encompass certain aspects of the climate change debate.

For more detailed background information, see CRS Report RL34042, *Environmental Services Markets: Farm Bill Proposals*, by Renée Johnson.

Considerations for Congress

Following is a list of questions that may be raised as Congress considers the role of the agriculture and forestry sectors as part of the broader climate change debate.

- **Farm Bill Programs.** Given the changes enacted in the 2008 farm bill, are there opportunities to expand existing federal conservation and land management programs to achieve greater emissions reduction and carbon sequestration in the agriculture sector? How might emissions reduction and carbon sequestration be integrated with the many other goals of conservation programs, such as improved soil quality and productivity, improved water and air quality, and wildlife habitat? Which programs or practices are the most beneficial and cost-effective? Are there ways to rank applications from farmers under existing programs to grant a higher weight to proposals to address climate change goals? Are there existing state programs that effectively address climate change and could be adopted at the federal level?

- **Emissions reductions.** Should carbon sequestration efforts be balanced by incentives to obtain additional emissions reductions in the agriculture sector through improved conservation and farm management practices, which could have a more immediate, direct, and lasting effect on overall GHG emissions? How might the existing regulatory framework for controlling air pollutants affect the climate change debate? What are the potential options for reducing GHG emissions at U.S. farming operations? How might cost concerns be addressed that limit broader adoption of manure management systems and also feed management strategies at U.S. livestock operations?

- **Carbon sequestration.** What are the upper limits of carbon capture and storage initiatives in the agriculture sector? For example, are such carbon sinks temporary or long-lasting, and what limits exist on their storage value? Do they rely appropriately on the willingness of landowners to adopt or continue to implement a particular conservation practice? Do they rely too heavily on the willingness of landowners to convert existing farmland to open space or prevent the conversion of existing farmland to non-farm uses? Are they cost-effective when compared to sinks in other sectors? How might concerns regarding uncertainty be addressed when measuring and estimating the amount of carbon sequestered in agricultural soils?

- **Carbon offset or credit markets.** What is the federal role in possibly expanding existing conservation programs in conjunction with efforts to create new market opportunities for farmers by developing a carbon credit trading system? How will USDA implement the new 2008 farm bill provision directing the Department to work with other agencies and organization to establish guidelines and standards for measuring agricultural and forestry environmental benefits, including carbon storage? What are the potential measurement, monitoring, enforcement, and administrative issues of implementing a carbon credit trading system involving the agriculture and forestry sectors? How would stored carbon be measured and verified; how much compensation would be available and for how long; whatare required management practices; and which accounting methodologies should be used? Would such a system operate under a voluntary or a mandatory framework?

- **Bioenergy promotion.** How might ongoing or anticipated initiatives to promote U.S. bioenergy production, such as corn-based or cellulosic ethanol, affect the options for land management or conservation strategies that could increase carbon uptake on agricultural lands and in agricultural soils? Might broader climate change goals be affected by increased agricultural production in response to corn-based ethanol? For example, might previously retired land be brought back into corn production or might this result in more intensive corn production, including fewer crop rotations and planting area setbacks, which could raise emissions and reduce the amount of carbon sequestered? Are there other competing commercial crops that might be used as a feedstock for ethanol that could also affect emissions and carbon uptake potential?

- **Energy efficiency.** What are the opportunities for improved on-farm energy efficiency and conservation? How might these be integrated into the broader framework on climate change mitigation in the agriculture sector?

- **Safeguarding U.S. agricultural production.** Among the possible effects of global climate change on agricultural production are increased climate variability and increased incidence of global environmental hazards, such as drought and/or flooding, pests, weeds, and diseases, or location shifts in where agriculture is produced. Climate change in some locations increases the yields of some crops. Some U.S. production regions are likely to fare better than others. Are additional initiatives needed in the U.S. agriculture sector to prepare for the potentially effects of global climate change that might impact U.S. agricultural production and food security? Which regions and crops might be "winners" or "losers" and how can transitions be eased?

APPENDIX: PRIMER ON AGRICULTURE'S ROLE IN THE CLIMATE CHANGE DEBATE

Question	Discussion
What are the types of GHG emissions assoc-iated with U.S. agriculture?	Official estimates of greenhouse gas (GHG) emissions for the U.S. agriculture sector are based on emissions of methane (CH_4) and nitrous oxide (N_2O) associated with agricultural production only. These estimates do not include carbon dioxide (CO_2) emissions from on-farm energy use and other emissions associated with forestry activities, food processing or distribution, or biofuel production. See **Agricultural GHG Emissions** in this report for more information.
What are the sources of GHG emissions from agriculture?	Agricultural sources of CH4 emissions are mostly associated with the natural digestive process of animals and with manure management on U.S. livestock operations. Sources of N_2O emissions are mostly associated with soil management and fertilizer use on U.S. croplands. **Figure 1** shows agricultural emissions by type and production category.
Why are CO_2 energy emissions excluded?	CO_2 emissions from on-farm energy use are aggregated with emissions for all transportation and industrial sectors, and comprise a small share of this total. Even if included in the estimates for the agriculture sector, this would not substantially raise agriculture's overall share of total GHG emissions.
What is agriculture's share of annual national GHG emissions?	In 2005, GHG emissions from U.S. agricultural activities totaled nearly 540 MMTCO_2-Eq (million metric tons CO_2-equivalent units, accounting for about 7% of annual national GHG emissions (**Table 1**). Fossil fuel combustion is the leading source of national GHG emissions (about 80%), with the energy sector generating about 85% of annual emissions across all U.S. sectors.
How much carbon is sequestered in U.S. agricultural soils?	In 2005, agricultural soils sequestered about 30 MMTCO_2-Eq., or roughly 5% of annual emissions generated from agricultural activities. Compared to total national GHG emissions, the agriculture sector offsets well under 1% of emissions annually. These estimates do not include uptake from forested lands or open areas that account for a majority (about 95%) of total U.S. sequestration. **Figure 2** shows carbon sequestration in agricultural soils. Also see **Agricultural Carbon Sinks** for more information.

(Continued)

Question	Discussion
Is there any uncertainty associated with estimates of carbon uptake for the agriculture sector?	Reasons for uncertainty associated with uptake estimates in U.S. soils include actual uptake depends on site specific conditions (e.g., location, climate, soil type, crop type, tillage practices, crop rotations, farm management, etc.); accounting methodology; type of practice, how well it is implemented, and the length of time undertaken; availability of federal/state cost-sharing or technical assistance; and other competing factors (including supply response for comercial crops and bioenergy crops). Actual GHG emissions may also vary according to many site-specific conditions. See **Uncertainty Estimating Carbon Sinks** for more information.
What is the potential to reduce emissions and/or increase carbon uptake in the agriculture sector?	The potential for carbon uptake in U.S. agriculture sector is much greater than current rates. USDA estimates net increases in carbon sequestration ranging from 40 to 590 MMTCO$_2$-Eq. per year (**Table 2**), or 2 to 20 times above current rates. This could offset total current national GHG emissions by as much as 8%. Other studies show an even greater carbon uptake potential in the agriculture sector. Practices that may reduce emissions and/or sequester carbon on U.S. farmlands include land retirement, pastureland and crop conversion, restoration; improved soil management and conservation tillage; and improved manure management and feeding strategies at livestock operations. See sections **Potential for Additional Uptake and Potential for Additional Reductions.**
How costly are the types of farming practices that help address climate change issues?	The estimated value (or cost) of sequestered carbon will vary by practice. USDA's forecast of an additional sequestration potential of 40 to 590 MMTCO$_2$-Eq. per year is associated with an estimated per-unit value ranging from \$3-\$34/mt of permanently sequestered carbon dioxide. The low-end of this range reflects the sequestration potential associated with cropland management practices; higher-end values are associated with afforestation and land retirement. See **Table 2** for more information See **Potential Mitigation Costs** for more information.
How can emissions from production be reduced? How can carbon uptake in	Most land management and agriculture conservation practices might both reduce GHG emissions and/or sequester carbon, including land retirement, conversion, and restoration; conservation tillage; soil management and

agricultural soils be increased?	soil erosion controls; efficient fertilizer/nutrient and chemical application; crop rotations; cover cropping; manure management; feed management; rotational grazing and improved forage; vegetative and riparian buffers; windbreaks for crops and livestock; bioenergy substitution and renewable energy use; and energy efficiency and energy conservation on-farm. See **Table 3** and **Mitigation Strategies in the Agriculture Sector** for more information.
Are there existing programs and/or legislation that promote farming practices that may help address climate change?	Existing federal and state farm conservation programs promote the types of land management and conservation practices that can reduce GHG emissions and/or sequester carbon. Also, many existing voluntary programs in the current farm bill, as well as under existing state-level programs provide cost-sharing and technical assistance to encourage farmers to implement such practices. These are voluntary programs and are generally designed to address site-specific improvements at an individual farming operation. See **Federal Programs** and other listed program information.

Source: Table prepared by the Congressional Research Service.

End Notes

[1] See CRS Report RL33849, *Climate Change: Science and Policy Options*, by Jane Leggett.

[2] See CRS Report RL34265, *Selected Issues Related to an Expansion of the Renewable Fuel Standard (RFS)*, by Brent D. Yacobucci and Randy Schnepf.

[3] EPA, *Inventory of U.S. Greenhouse Gas Emissions and Sinks: 1990-2005*, April 2007, at [http://epa.gov/climatechange/emissions/usinventoryreport.html].

[4] USDA, *U.S. Agriculture and Forestry Greenhouse Gas Inventory: 1990-2001*, TB1907, March 2004, at [http://www.usda.gov/oce/global_change/gg_inventory.htm].

[5] "Carbon-equivalents" equate an amount of a GHG to the amount of carbon that could have a similar impact on global temperature. EPA's data are in teragrams (million metric tons). Alternative ways to express emissions and offsets are in carbon equivalents (MMTCE), which assume a multiplier of 0.272 to convert fromEPA-reported equivalent CO_2-Eq. units.

[6] The principal gases associated with climate change from human activities are CO_2, CH_4, N_2O, and ozone-depleting substances and chlorinated and fluorinated gases, such as hydrofluorocarbons, perfluorocarbons, and sulfur hexafluoride. See CRS Report RL33849, *Climate Change: Science and Policy Implications*, by Jane Leggett.

[7] Methane's ability to trap heat in the atmosphere is 21 times that of CO_2; nitrous oxide is 310 times that of CO_2 (measured over a 100-year period). Intergovernmental Panel on Climate Change (IPCC), *Climate Change 2007,* Technical Summary of the Working Group I Report, Table TS-2, at [http://ipcc-wg1.ucar.edu/wg1/Report/AR4WG1_Print_TS.pdf].

[8] USDA, *U.S. Agriculture and Forestry Greenhouse Gas Inventory: 1990-2001*, TB1907, Figure 3-6, March 2004, at [http://www.usda.gov/oce/global_change/gg_inventory.htm]. Nitrogen-fixing crops refer to beans, legumes, alfalfa, and non-alfalfa forage crops.

[9] EPA's 2007 *Inventory,* Table ES-2. Other major CH_4 sources were landfills, natural gas systems, and coal mining. Mobile combustion was the second largest source of N_2O.

[10] EPA's 2007 *Inventory,* Table 6-2. NO_X and CO influence the levels of tropospheric ozone, which is both a local pollutantand a GHG (called "indirect" greenhouse gases). Their contributions cannot be measured by emissions.

[11] See CRS Report RL32948, *Air Quality Issues and Animal Agriculture: A Primer*, byClaudia Copeland. Particulate emissions may also contribute to climate change, but theirinfluence is predominantly local, short-term and poorly quantified.

[12] EPA's 2007 *Inventory,* Table 2-14 and Table 6-1.

[13] Aside from the energy and agriculture/forestry sectors, by source, other leading contributors are wood biomass/ethanol use (3%); nonenergy use of fuel; landfills; and substitution of ozone-depleting substances (2% each). By sector, leading sources are industrial processes (5%) and wastes (2%). EPA's 2007 *Inventory,* Tables ES-2 and ES-4.

[14] The IPCC was established to assess scientific, technical and socioeconomic information related to climate change, its potential impacts and options for adaptation and mitigation. IPCC's methodolgy to estimate emissions and sinks are consistent with those used by other governments and with established guidelines under the United Nations Framework Convention on Climate Change.

[15] Land use and forestry activities account for less than 1% of total estimated GHG emissions in the United States (EPA's 2007 *Inventory,* Table ES-4). See Table 1.

[16] EPA's 2007 *Inventory,* Table 2-14.

[17] Refers to livestock (cattle, sheep, goats, and buffalo) that have a four-chambered stomach. In the rumen chamber, bacteria breaks down food and degrades methane as a byproduct.

[18] R. A. Leng, "Quantitative Ruminant Nutrition — A Green Science," *Australian Journal of Agricultural Research*, 44: 363-380. Feed efficiency based on both fermentive digestion in the rumen and conversion of feed to output (e.,g, milk, meat) as nutrients are absorbed.

[19] Although carbon is released as well, it is predominantly absorbed again within a year as part of the cropping cycle, and so is assumed to be net zero emissions unless some goes into long-term soil carbon content.

[20] Previously estimates for the agriculture soil management category were lower. Current EPA estimates reflect methodological and input data changes.

[21] D. C. Reicosky, "Environmental Benefits of Soil Carbon Sequestration," USDA, at [http://www.dep.state.pa.us/dep/DEPUTATE/Watermgt/wsm/WSM_TAO/InnovTechForum/InnovTechForum-IIE-Reicosky.pdf].

[22] USDA, "Conservation Tillage Firmly Planted in U.S. Agriculture," *Agricultural Outlook*, March 2001; USDA, "To Plow or Not to Plow? Balancing Slug Populations With Environmental Concerns and Soil Health," *Agricultural Research,* October 2004; Conservation Technology Information Center (CTIC), "Conservation Tillage Facts," at [http://www.conservationinformation.org/?action=learningcenter_core4_convotill].

[23] An enclosed tank that promotes decomposition using anaerobic conditions and naturally occurring bacteria, while producing biogas as a byproduct that can be used as energy.

[24] See CRS Report RL32948, *Air Quality Issues and Animal Agriculture: A Primer*; and CRS Report RL31851, *Animal Waste and Water Quality: EPA Regulation of Concentrated Animal Feeding Operations (CAFOs)*, by Claudia Copeland.

[25] R. Pillars, "Farm-based Anaerobic Digesters," Michigan State University Extension, at [http://web2.msue.msu.edu/manure/FinalAnearobicDigestionFactsheet.pdf].

[26] EPA, *Development Document for the Final Revisions to the NPDES Regulation and the Effluent Guidelines for Concentrated Animal Feeding Operations*, January 2003.

[27] C. Henry and R. Koelsch, "What Is an Anaerobic Digester?" University of Nebraska, Lincoln, at [http://files.harc.edu/Sites/GulfcoastCHP/Publications/WhatIsAnaerobic Digestion.pdf]; and

Pennsylvania State University, "Biogas and Anaerobic Digestion," at [http://www.biogas.psu.edu/]. For optimum operation, anaerobic digesters must be kept at a constant, elevated temperature, and any rapid changes in temperature could disrupt bacterial activity.

28 Mostly Section 9006 and Section 6013 of the farm bill (P.L. 107-171), but also under other farm bill cost-share programs. CRS communication with USDA staff.

29 As of 2005. EPA, *AgStar Digest*, Winter 2006, at [http://www.epa.gov/agstar/].

30 R. A. Leng, "Quantitative Ruminant Nutrition — A Green Science," *Australian Journal of Agricultural Research*, 44: 363-380; H. Steinfeld, C. de Haan, and H. Blackburn, *Livestock-Environment Interactions, Issues and Options*, chapter 3 (study commissioned by the Commission of the European Communities, United Nations, and World Bank), at [http://www.virtualcentre.org/es/dec/toolbox/FAO/Summary/index.htm].

31 U.S. Geological Survey (USGS), website information on carbon sequestration in soils.

32 EPA's 2007 *Inventory*, Table 2-14 and Table 7-1. Based on estimates for the following categories: land converted to grassland; grassland remaining grassland; land converted to cropland; cropland remaining cropland.

33 Congressional Budget Office (CBO), *The Potential for Carbon Sequestration in the United States*, Sept. 2007, at [http://www.cbo.gov/ftpdocs/86xx/doc8624/09-12-Carbon Sequestration.pdf]. Biological sequestration refers to the use of land to enhance its ability to uptake carbon from atmosphere through plants and soils. Direct sequestration refers to capturing carbon at its source and storing it before its release to the atmosphere. Examples include capture and storage in geologic formations, such as oil fields, natural gas fields, coal seams, and deep saline formations. See CRS Report RL33801, *Carbon Capture and Sequestration (CCS)*, by Peter Folger.

34 EPA's 2007 *Inventory*, Table 2-14 and Table 7-1. Based on estimates for the following categories: forestland remaining forestland; and growth in urban trees. Other uptake not included in the estimates is from landfilled yard trimmings.

35 See, for example, T. A. Butt and B. A. McCarl, "Implications of Carbon Sequestration for Landowners," *2005 Journal of the American Society of Farm Managers and Rural Appraisers*; Government Accountability Office (GAO), *Conservation Reserve Program: Cost-Effectiveness Is Uncertain*, March 1993; H. Feng, J.Zhao, and C. Kling, "Carbon: The Next Big Cash Crop," *Choices*, 2nd quarter 2001; and H. Feng, C. Kling, and P. Glassman, "Carbon Sequestration, Co-Benefits, and Conservation Programs," *Choices*, Fall 2004.

36 W. Hohenstein, "USDA Activities to Address Greenhouse Gases and Carbon Sequestration," presentation to Senate Energy Committee staff, February 15, 2007.

37 USDA, *Economics of Sequestering Carbon in the U.S. Agricultural Sector*, April 2004.

38 EPA, *Greenhouse Gas Mitigation Potential in U.S. Forestry and Agriculture*, Tables 4-10 and 4-5, Nov. 2005, at [http://www.epa.gov/sequestration/greenhouse_gas.html].

39 As summarized by CBO, *The Potential for Carbon Sequestration in the United States*, Sept. 2007, at [http://www.cbo.gov/ftpdocs/86xx/doc8624/09-12-CarbonSequestration.pdf].

40 USDA, *Economics of Sequestering Carbon in the U.S. Agricultural Sector*, April 2004 (measured by the amount of carbon sequestered over a 15-year time period across a range of costs). USDA estimates that the associated total cost to sequester carbon across this range is $0.95 billion to $2 billion per year.

41 EPA, *Greenhouse Gas Mitigation Potential in U.S. Forestry and Agriculture*, Table 4-10.

42 USDA, *U.S. Agriculture and Forestry Greenhouse Gas Inventory: 1990-2001*, TB1907, Figure 3-8, March 2004, at [http://www.usda.gov/oce/global_change/gg_inventory.htm].

43 Bongen, A.,"Using Agricultural Land for Carbon Sequestration," Purdue University, at [http://www.agry.purdue.edu/soils/Csequest.PDF]. 1999 data for carbon storage inIndiana.

44 USDA, *U.S. Agriculture and Forestry Greenhouse Gas Inventory: 1990-2001*, TB1907, March 2004, at [http://www.usda.gov/oce/global_change/gg_inventory.htm]; USDA, "Depositing Carbon in the Bank: The Soil Bank, That Is," *Agricultural Research*, Feb. 2001.

45 See Iowa Farm Bureau's carbon credit project at [http://www.iowafarmbureau.com].

[46] USDA, "Reducing Agricultural Greenhouse Gas Emissions Through Voluntary Action," Statement by Bruce Knight of USDA's Natural Resources Conservation Service at the United Nations Framework Convention on Climate Change, December 2004, at [http://www.nrcs.usda.gov/news/speeches04/climatechange.html]

[47] USDA, "USDA Targeted Incentives for Greenhouse Gas Sequestration," June 6, 2003; W. Hohenstein, "USDA Conservation Programs are Targeting Greenhouse Gases and Carbon Sequestration." Provided to Senate Energy Committee staff, February 15, 2007.

[48] A previous program in Title VI (Rural Development) that was not reauthorized in the 2008 farm bill was a provision (Section 6013) authorizing rural development business and industry program to make loans and loan guarantees for renewable energy systems, including wind energy systems and anaerobic digesters.

[49] Previously referred to as Section 9006 (Renewable Energy Systems and Energy Efficiency Improvements) in the 2002 farm bill.

[50] CRS communication with USDA staff, February 8, 2007. Limited information indicates that USDA funded eight projects totaling more than $60 million under the previous Section 6013 and provided another $20 million in funding assistance under Section 9006 for anaerobic digesters (FY2002-FY2005).

[51] See CRS Report RL34130, *Renewable Energy Policy in the 2007 Farm Bill*, and CRS Report RL32712, *Agriculture-Based Renewable Energy Production*, both by Randy Schnepf; and CRS Report RL33572, *Biofuels Incentives: A Summary of Federal Programs*, by Brent Yacobucci.

[52] A previous program that was not reauthorized in the 2008 farm bill was the Forest Service's Forest Land Enhancement Program (FLEP). FLEP provided funding for agriculture and silvopasture practices with rotational grazing and improved forage. Primary efforts under the program included afforestation and reforestation, improved forest stand, constructing windbreaks, and riparian forest buffers. For information on USDA forestry programs, see CRS Report RL33917, *Forestry in the 2008 Farm Bill*, by Ross W. Gorte.

[53] State and Local Government directory at [http://www.statelocalgov.net/index.cfm].

[54] Pew Center, *Learning from State Action on Climate Change*, Oct. 2006, [http://www.pewclimate.org/policy_center/policy_reports_and_analysis/state].

[55] See CRS Report RL33812, *Climate Change: Actions by States to Address Greenhouse Gas Emissions*, by Jonathan L. Ramseur.

[56] California's Global Warming Solutions Act of 2006 (AB 32), which was enacted in September 2006, codified the state's goal of requiring California's GHG emissions be reduced to 1990 levels by 2020.

[57] California Environmental Protection Agency, *Expanded List of Early Action Measures to Reduce Greenhouse Gas Emissions in California Recommended for Board Consideration*, Oct. 2007, at [http://www.arb.ca.gov/cc/ccea/meetings/ea_final_report.pdf].

[58] California EPA, "Expanded List of Early Action Measures to Reduce Greenhouse Gas Emissions in California Recommended for Board Consideration," October 2007, at [http://www.arb.ca.gov/cc/ccea/meetings/ea_final_report.pdf].

[59] For additional general information on voluntary carbon markets, see CRS Report RL34241, *Voluntary Carbon Offsets: Overview and Assessment*, by Jonathan L. Ramseur. For tradingpurposes, one carbon credit is considered equivalent to one metric ton of carbon dioxide emission reduced.

[60] Iowa Farm Bureau, Carbon Credit Aggregation Pilot Project, at [http://www. iowafarmbureau.com/special/carbon/]; CRS staff communication with Iowa Farm Bureau staff, January 2007.

[61] The Exchange is a voluntary, self-regulated, rules-based exchange. Its emission offset program constitutes a small part of its overall program, which includes methane destruction, carbon sequestration, and renewable energy. See [http://www.chicagoclimatex.com/].

[62] For more information, see North Dakota Farmers Union [http://www.ndfu.org], Illinois Conservation and Climate Initiative [http://www.illinoisclimate.org], Environmental Credit Corporation [http://www.envcc.com]; and Terrapass [http://www.terrapass.com/projects].

[63] CRS estimate based on information from the Iowa Farm Bureau (January 17, 2008).

[64] Other similar programs include the Illinois Conservation and Climate Initiative, the Environmental Credit Corporation (Indiana), the Upper Columbia Resource Conservation and Development Council (Northwest), and Terrapass (California), among others.

[65] Committee on Energy and Commerce, "Climate ChangeLegislation Design White paper: Scope of a Cap-and-Trade Program," prepared by committee staff, Oct. 2007, available at [http://energycommerce.house.gov/Climate_Change/White_Paper.100307.pdf].

[66] Including H.R. 1590 (Waxman), S. 309 (Sanders), and S. 485 (Kerry).

[67] See, for example, a study conducted for the Fertilizer Institute at [http://www.tfi.org/issues/climate/doanestudy.pdf].

[68]GHG bills that provide for agriculture or forestry offsets are S. 2191 (Lieberman/Warner), S. 280 (McCain/Lieberman), S. 317 (Feinstein), S. 1168 (Alexander/Lieberman), S. 1177 (Carper), S. 1766 (Bingaman/Specter), and H.R. 620 (Olver). Markey (H.R. 6186) would allow for agriculture offsets, except from animal operations under performance standards.

[69] Primarily S. 2191 and also S. 1766 (Bingaman/Specter).

In the context of credit trading, an offset is a certificate representing the reduction of one metric ton of carbon dioxide emissions, the principal greenhouse gas. Offsets generally fall within the categories of biological sequestration, renewable energy, energy efficiency, and reduction of non-CO_2 emissions.

[71] For more information on allowances and auction proceeds in current GHG bills, see *Allocations for Carbon Allowances and Auctions under S. 2191*, by Brent D. Yacobucci, CRS general distribution memorandum, February 22, 2008.

[72] Committee on Energy and Commerce, "Climate ChangeLegislation DesignWhite paper: Scope of a Cap-and-Trade Program," prepared by committee staff, Oct. 2007, available at [http://energycommerce.house.gov/Climate_Change/White_Paper.100307.pdf].

[73] National Association of Wheat Growers, "Ag, Utility Groups Write on Stabenow Amendment," June 13, 2008, at [http://www.wheatworld.org/html/news.cfm?ID=1423].

[74] Senators Bob Dole and Tom Daschle, *The Role of Agriculture in Reducing Greenhouse Gas Emissions: Recommendations for a National Cap-and-Trade Program*, April 2008, at [http://www.bipartisanpolicy.org/ht/display/ArticleDetails/i/6086].

[75] See, for example, E. Boyd, E. Corbera, B. Kjellén, M. Guitiérrez, and M. Estrada, "The Politics of 'Sinks' and the CDM: A Process Tracing of the UNFCCC Negotiations(pre-Kyoto to COP-9)," Feb. 2007, draft submitted for International EnvironmentalAgreements; also see two articles in *Nature*, no. 6812, Nov. 2000, "Deadlock in the Hague,but Hope Remains for Spring Climate Deal," and "Critical Politics of Carbon Sinks."

[76] Referred to as "land use, land use change, forestry," or abbreviated as LULUCF.

[77] Subcommittee on Rural Revitalization, Conservation, Forestry and Credit hearing, May 21, 2008, "Creating Jobs with Climate Solutions: How agriculture and forestry can help lower costs in a low-carbon economy," at [http://agriculture.senate.gov/].

In: The Role of Agriculture in Carbon Capture... ISBN: 978-1-60741-445-2
Editor: Arvid Bjurstrom

Chapter 3

MEASURING AND MONITORING CARBON IN THE AGRICULTURAL AND FORESTRY SECTORS*

[1]Ross W. Gorte and [2]Renée Johnson
[1] Natural Resources Policy
[2] Agriculture Policy Resources, Science and Industry Division

SUMMARY

Proposals to reduce emissions of carbon dioxide and other greenhouse gases often include the use of forestry and agricultural practices and lands for carbon sequestration. However, uncertainty about the accuracy of measuring carbon from these activities has led some to question this potential. Basic approaches for measuring forest and agricultural carbon include on-site measurement; indirect measurement from off-site tools; and estimation using models or inferences. Because of challenges associated with balancing the cost and accuracy of these measurement tools, any practicable system for measuring forest and agricultural carbon might require a mix of these approaches.

Concerns about global climate change and its impacts on the environment and the economy are encouraging policy-makers and stakeholders to explore a range

---

* This is an edited, reformatted and augmented version of a CRS Report for Congress publication dated October 2008.

of options to reduce emissions of carbon dioxide (CO_2) and other greenhouse gases (GHGs).[1] Congress is considering legislation that would, among other things, provide incentives for parties to reduce or mitigate GHG emissions or to sequester (store) additional CO_2.[2] The possible use of forestry and agricultural practices and lands to mitigate or sequester CO_2 is part of the debate. However, substantial uncertainty exists about current ability to accurately quantify, monitor, and verify the amount of carbon sequestered by various agricultural and forestry practices. By comparison, measuring the carbon from a discrete point source, such as a power plant, is relatively easy and precise. Incorporating the agriculture and forestry sectors in an emissions reduction program will likely require a firm basis for measuring carbon inventories and change for forestry and agricultural practices and lands.

PURPOSE OF MEASURING FOREST AND AGRICULTURAL CARBON

Farm and forest activities can be both a source and a sink of GHGs, releasing GHGs through plant and animal respiration and decomposition and removing CO_2 through photosynthesis, storing it in vegetation and soils (a process known as sequestration). A range of land management, agricultural conservation, and other farmland practices can reduce or abate emissions and/or sequester carbon. These include tree planting, soil conservation, manure and grazing management, and land retirement, conversion, and restoration.[3] Many of these activities, however, may be impracticable for an emission trading program because they might not meet credible standards for quantifying, monitoring, and verifying emission reduction or carbon storage.

Reliable tools and techniques are needed for carbon inventories and carbon change on forests and agricultural lands. The ability to measure carbon levels allows countries that have committed to reducingGHG emissions to measure their current annual emissions and carbon storage (and changes in carbon stocks).[4] Current estimates show that forests account for a significant share of estimated existing carbon stocks globally; agricultural lands account for a small share of stored carbon. Also, the ability to measure carbon levels provides the means to estimate the mitigation potential of forest or agriculture activities that sequester additional carbon in soils or vegetation (i.e., result in a net reduction compared to estimated baseline conditions or current sequestration). This may allow a farm or forestry activity to be recognized as a way to mitigate or *offset*[5] emissions — through voluntary action, an emissions trading market, or a regulatory program.

For an emissions trading program to be credible, a participating entity is usually required to meet a series of established protocols that specify what, when, where, and how to measure changes in carbon. Protocols provide technical guidelines or standardized rules for quantifying, monitoring, and verifying the mitigation of an activity. They specify requirements on project eligibility, scale and baseline measurements, measurement frequency, and verification. The difficulty is developing credible protocols that are quantitatively defensible and readily applicable across areas with differing land uses, weather, and other site-specific conditions. Protocols also address, to varying degrees, concerns about the validity of activities, such as additionality, leakage, and permanence.[6]

Protocols may be either voluntary or set by regulation. In one voluntary market, the Chicago Climate Exchange (CCX) has protocols for a range of soil and land management projects, including agricultural methane, soil carbon, rangeland soil carbon management, and tree planting projects.[7] The Regional Greenhouse Gas Initiative (RGGI) — the first regional mandatory, market-based effort to reduce GHG emissions — is developing technical standards for a narrower set of offset projects from the agricultural and forestry sectors, providing for afforestation and methane reduction from livestock operations.[8] Individual requirements of current protocols and standards can vary widely by program.

Decisions Needed in Setting Measurement Requirements

Numerous methods exist to measure forest and agricultural carbon. The appropriate measure to use in specific circumstances depends on several variables, including the purpose for measuring the carbon, the scale and basis to be measured, the frequency of the measurement, and how the measurement is to be verified.

Scale and Baseline

Two geographic scales are commonly used for measuring GHG emissions — the national/regional level to report GHG emissions and participate in broad efforts to reduce emissions; and the local/site-specific level for projects to offset emissions. Regardless of scale, the emission reduction or carbon sequestration is compared to a *baseline* — the historic GHG emissions or carbon stocks at a specified point in time. The scale and baseline timing are typically specified in the protocol of the reporting, marketing, or regulating organization. Sometimes, for projects with multiple land uses, the land is stratified into the various land uses (e.g., cropland, pasture, sapling forest, mature forest), with a different baseline established for each use.[9]

Periodicity

Protocols typically identify when GHG emissions must be measured. An initial measurement is needed to establish the baseline. This must be done prior to the onset of a project, to allow for measuring the *change* that results from the action. Occasionally, a historic baseline is specified; for example, the Kyoto Protocol identified 1990 emissions as the baseline for measuring emission reductions. Other options include a current level, or other level whereby a project is compared to "business as usual." The protocols also identify the frequency and timing of measurements. For example, CCX contracts for agricultural projects require annual measurements to assure that the emission reduction or carbon sequestration is actually occurring.

Frequency of measurement also depends on the rate of change in carbon storage. Some carbon pools, such as forest soils, change relatively slowly (unless the forest is disturbed), and measurement once a decade may be sufficient. For other carbon pools, such as pastures or managed lands, differences within and across years can be substantial, and may require more frequent measurement. Timing can be critical, and alternative measurements may vary widely. The amount of carbon stored in vegetation, in particular, varies over the course of a year, with carbon being sequestered during the spring, carbon stored in foliage at its maximum in late summer, and carbon being released during the winter as the deciduous leaves decompose. Thus, consistent timing for annual measures is an important element for agricultural and forestry carbon projects.

Verification

Verifying the emission reduction or carbon sequestration is critical in efforts to mitigate climate change. It is particularly important for agriculture and forestry projects, as theseactivities areharder to measure reliably than other types of GHG offsets. One question is who will be responsible for verifying changes in carbon, which raises questions about the role of a regulatory agency for accrediting claimed changes in carbon levels from an activity.

Existing programs typically recommend or require that the carbon offset be verified by an independent entity. Independent verification may be an auditing function, to assure the reality and accuracy of the carbon offset for markets (buyers and sellers), regulations (emitters and regulators), and reports (emitters and reporting organizations).[10] One source has prescribed several qualities for independent verification: an "independent, qualified, third-party verifier" using "approved methodologies and regulations" and "whose compensation is not in any way dependent on the outcomes of their decisions" and who follows set procedures to avoid conflicts of interest.[11]

As voluntary and regulated markets for GHG emissions offsets develop, qualified, independent organizations to verify carbon offsets will be needed. Entities qualified to verify agriculture and forest carbon offsets must be proven to be knowledgeable about carbon measurement. One source notes: "To provide good quality and trustworthy oversight, a sufficiently rigorous accreditation process will be necessary to ensure that the verifiers have the needed expertise."[12] This process could parallel the development of independent auditors for certifying sustainable forestry programs.[13]

Measurement Techniques

Basic approaches for measuring agricultural and forest carbon inventories and change include on-site measurement, indirect measurement from off-site tools, and estimation using process models or inferences. A hybrid approach involving a combination of approaches (e.g., combining modeling with on-site sampling and independent verification) might improve the accuracy enough to be useful while still containing costs. Because of the inherent challenges associated with balancing the cost of measuring carbon and the accuracy of these measurements, any practicable system for measuring forest and agricultural carbon might require a mix of these different methods and approaches, rather than a single approach.

On-Site Measurement

Direct measurement of the carbon content of agricultural and forestry soils and vegetation through field sampling and site-specific laboratory estimates is perhaps the most accurate way to measure carbon levels and changes. However, this is time-consuming, costly, and often requires continuous sampling and replication via a census of soil and vegetation carbon for all agriculture and forestry projects, and may be infeasible. Also, it cannot cover large areas. Samples can be taken and the results extrapolated, based on soil survey, land cover, climate, and other spatial data. Sampling patterns (e.g., a grid, random, or stratified random), intensity (e.g., the area to be sampled), and frequency are likely to be specified in the protocols, and many sources discuss sampling methods for agriculture and forestry projects.[14] The more intensive and frequent the sampling, the greater the cost, but the higher the likely accuracy of the data. Most experts suggest some sampling to ensure the accuracy of models or off-site measures, especially performed consistently over time.

As with verification, the entity that measures the on-site carbon can affect perceptions of the accuracy of the measurement. Landowners or other offset sellers can perform the measurement — both at the outset of the project (for the baseline) and periodically during the life of the project. This could reduce costs, because they are commonly on the site, but raises questions of credibility, since they have an interest in the reported carbon levels. Ensuring that verification is conducted by independent verifiers might be sufficient to assuage market concerns over credibility, but could involve high project verification costs.

Indirect Measurement with Off-Site Tools

Tools exist to calculate carbon content without actually being on the site. Remote sensing — using photographic and other images from aircraft or satellites — can be used to measure carbon-related factors. For example, infrared imagery can detect live biomass, with variations in the image reflecting variations in the type and level of biomass. Remote sensing has long been used in forestry for calculating commercial timber volumes of forest stands.

The principal advantage of remote sensing is coverage, given its ability to assess a wide area relativelyquickly. Another advantage isthat the remote sensing and the analysis of the results are generally performed by experts, improving the credibility of the results and probably lowering the cost of verification. It can provide highly accurate information for some types of carbon-related measures, such as activities with readily visible results (e.g., deforestation and afforestation) or measurable carbon pools (e.g., live above-ground biomass). One disadvantage is the high fixed cost of providing remote coverage; satellites are very expensive to launch and maintain. Aircraft may be less expensive but may cover less area. Once the satellites are in place, extending satellite coverage to additional areas is relatively inexpensive. For some carbon-related measures, such as activities with less visible impacts (e.g., sustainable forestry) or less readily measurable carbon pools (e.g., soil carbon), remote sensing is problematic. Also, in some areas, cloud cover can interrupt regular measurements. Methods for consistently and reliably interpreting remote imagery are still under development, and are usually recommended to be used in conjunction with other techniques.

Estimation Using Process Models or Inferences

Another indirect approach is to estimate agricultural and forestry carbon with models or other analytical tools. Models are available to estimate a variety of ecosystem processes, and are used to depict site-specific conditions. Models, especially computer models, are typically built from extensive research and datasets, and provide average or archetypical estimates of physical area, temperature, precipitation, forest or soil type, slope, plant diversity, and microbial activity. The accuracy of the results depends in large part on the validity and measurement of the input variables for the model. Data may be presented in tabular form, called "look-up tables" because estimates can be looked up in the table based on a few key variables, such as forest type and tree age or soil type.[15] A related simpler approach might use inferences or generalized input data scaled up to the size of the farm or forested area to approximate the sequestration for an activity.[16] Such an approach may reduce costs, but provide a relatively low level of precision, and possibly high verification costs.

The advantage of a modeling approach is thatit is relatively simple and low-cost, and often provides consistent estimates. However, it may not reflect actual differences within and across sites and activities, since it relies on archetypical or average carbon estimates and not site-specific carbon measurements. Model proponents often suggest using occasional site-specific sampling to assure the validity of the model chosen for the project and site, and some suggest adjusting the estimates based on the samples. This introduces the potential for bias in reporting carbon, and significantly increases the difficulty of verification. In addition, for most situations and project types, numerous models exist. These competing models may yield quite different estimates for the same site, because of the different data sets and assumptions used in constructing the models. One model may yield the most accurate estimates in certain circumstances, while another model may yield more accurate estimates in other circumstances.

Considerations for Congress

Congress has already taken steps to address some of the challenges associated with measuring carbon changes from forested and agricultural lands and practices. The 2008 farm bill (P.L. 110-246, the Food, Conservation, and Energy Act of 2008) includes a provision (Sec. 2709) directing USDA to "establish technical guidelines that outline science-based methods to measure the environmental servicesbenefits," including carbon storage, from forests and agricultural activities. This includes developing measurement procedures and a reporting protocol and registry. The Energy Independence and Security Act of 2007 (P.L. 110-140, Sec. 712) directs the Secretary of the Interior to develop a methodology to assess carbon sequestration and emissions from ecosystems. This methodology is to cover measuring, monitoring, and quantifying GHG emissions and reductions, and provide estimates of sequestration capacity and the mitigation potential of different ecosystem management practices.

Congress continues to face the question of whether its current authorized activities provide adequate and sufficient guidelines for accurately measuring carbon levels from forest and agricultural activities.

End Notes

[1] Under the United NationsFrameworkConvention on Climate Change (UNFCC), GHGs include CO_2, methane (CH_4), nitrous oxide (N_2O), hydrofluorocarbons (HFCs), perfluorocarbons (PFCs), and sulfur hexafluoride (SF_6). Because various GHGs have different climatic consequences, they are typicallyconverted to a standard measure, usually metric tons of CO_2-equivalents (CO_2-Eq.).

[2] CRS Report RL33846, *Greenhouse Gas Reduction: Cap-and-Trade Bills in the 110th Congress.*

[3] See CRS Report RL33898, *Climate Change: The Role of the U.S. Agriculture Sector.*

[4] The official U.S. estimates of current national GHG emissions and carbon uptake, including agriculture and forestry estimates, are those published by the U.S. Environmental Protection Agency (EPA) in its annual *Inventory of U.S. Greenhouse Gas Emissions and Sinks.*

[5] In this report, *offset* refers to any action that reduces or mitigates GHG emissions, usually from an unrelated source (e.g., increased carbon storage on forest or farmlands to offset emissions from automobiles). The term *offsets* may also be used to refer to approved carbon reduction or sequestration projects under specific regulatory or voluntary GHG reduction programs. See CRS Report RL34560, *Forest Carbon Markets: Potential and Drawbacks.*

[6] See CRS Report RL34436, *The Role of Offsets in a Greenhouse Gas Emissions Cap-and-Trade Program: Potential Benefits and Concerns*; and Timothy Pearson et al., *Sourcebook for Land Use, Land-Use Change, and Forestry Projects*, Winrock International, 2005.

[7] CCX, "CCX Offsets Program," at [http://www.chicagoclimatex.com/content.jsf?id=23].

[8] RGGI, *Regional Greenhouse Gas Initiative Model Rule,* 1/5/07 Final, at [http://rggi.org/docs/model_rule_corrected_1_5_07.pdf].

[9] See Suzie Greenhalgh et al., *The Greenhouse Gas Protocol: The Land Use, Land-Use Change, and Forestry Guidance for GHG Project Accounting*, World Resources Institute, Oct. 2006.
[10] Zach Willey and Bill Chameides, eds., *Harnessing Farms and Forests in the Low-Carbon Economy: How to Create, Measure, and Verify Greenhouse Gas Offsets,* Nicholas Institute for Environmental Policy Solutions, 2007.
[11] Offset Quality Initiative, *Ensuring Offset Quality* (July 2008), at [http://www.pewclimate.org/].
[12] Lydia Olander et al., *Designing Offsets Policy for the U.S.: Principles, Challenges, and Options for Encouraging Domestic and International Emissions Reductions and Sequestration from Uncapped Entities as Part of a Federal Cap-and-Trade for Greenhouse Gases,* Nicholas Institute for Environmental Policy Solutions, May 2008.
[13] For more information on forest certification, see [http://www.pinchot.org/project/59].
[14] For examples of the latter, see *Harnessing Forest and Farm Carbon; GHG Project Accounting; and Sourcebook for LULUCF Projects.*
[15] James E. Smith et al., *Methods for Calculating Forest Ecosystem and Harvested Carbon with Standard Estimates for Forest Types of the United States*, Gen. Tech. Rept. NE-343, April 2006.
[16] See, e.g., U.S. Dept. of Energy, *Technical Guidelines Voluntary Reporting of Greenhouse Gases (1605(b)) Program,* March 2006, Parts H and I.

INDEX

A

B

C

D

M

N

O

P

R

S

Y